HOP INTO ACTION

The Amphibian Curriculum Guide for Grades K–4

HOP INTO ACTION

The Amphibian Curriculum Guide for Grades K–4

David
Alexander

NSTApress
National Science Teachers Association

Arlington, Virginia

Claire Reinburg, Director
Jennifer Horak, Managing Editor
Andrew Cooke, Senior Editor
Judy Cusick, Senior Editor
Wendy Rubin, Associate Editor
Amy America, Book Acquisitions Coordinator

ART AND DESIGN
Will Thomas Jr., Director
Joe Butera, Senior Graphic Designer, cover and interior design
Original art contributed by Natalia Hubisz
Cover illustrations by Laurent Renault and Rorat for iStock

PRINTING AND PRODUCTION
Catherine Lorrain, Director
Nguyet Tran, Assistant Production Manager

NATIONAL SCIENCE TEACHERS ASSOCIATION
Francis Q. Eberle, PhD, Executive Director
David Beacom, Publisher

Library of Congress Cataloging-in-Publication Data

Alexander, David, 1983-
  Hop into action: the curriculum guide for grades K4/by David Alexander.
     p. cm.
  Includes index.
  ISBN 978-1-936137-07-7
  1. Amphibians—Study and teaching (Elementary) I. Title.
  QL645.6.A45 2010
  372.35'7—dc22

                              2010032137

  eISBN 978-1-936137-57-2

NSTA is committed to publishing material that promotes the best in inquiry-based science education. However, conditions of actual use may vary, and the safety procedures and practices described in this book are intended to serve only as a guide. Additional precautionary measures may be required. NSTA and the authors do not warrant or represent that the procedures and practices in this book meet any safety code or standard of federal, state, or local regulations. NSTA and the authors disclaim any liability for personal injury or damage to property arising out of or relating to the use of this book, including any of the recommendations, instructions, or materials contained therein.

PERMISSIONS
You may photocopy, print, or e-mail up to five copies of an NSTA book chapter for personal use only; this does not include display or promotional use. Elementary, middle, and high school teachers *only* may reproduce a single NSTA book chapter for classroom or noncommercial, professional-development use only. For permission to photocopy or use material electronically from this NSTA Press book, please contact the Copyright Clearance Center (CCC) (*www.copyright.com*; 978-750-8400). Please access *www.nsta.org/permissions* for further information about NSTA's rights and permissions policies.

CONTENTS

Amphibian Education Lessons

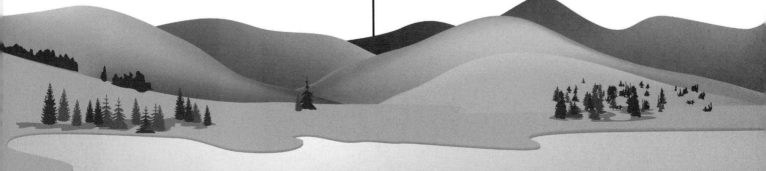

Preface

Amphibian education and community involvement is critical at this time. We face a global amphibian extinction crisis. Of the more than 6,000 recognized species of amphibians, more than one third are suffering serious declines or have recently become extinct, despite having survived millions of years. If we do not educate our youth to appreciate, understand, and take action for amphibians and their environments, the amphibians are destined to go the way of the dinosaurs. The *Hop Into Action* curriculum guide was developed in response to this urgency in order to arm educators from a variety of settings with tools they can use to incorporate effective environmental education for learners in kindergarten through fourth grade.

As an educator, you are the audience for this guide, which offers 20 lessons that can be used individually or as a curriculum. In addition it

- includes interdisciplinary approaches to curriculum areas to meet national standards;
- is designed for classroom teachers, home school educators, naturalists, and camp leaders;
- provides lessons geared for grades K–4, with some appropriate content and extensions for younger and older grade levels; and
- was created from the firsthand experience of educators in both formal and informal learning environments.

Because amphibians form a link between aquatic and terrestrial environments, they offer exciting opportunities for education and also can be used to educate across multiple subject areas. Educating students about these species will be critical to foster awareness and social concern that may one day lead to stewardship and conservation. Lessons provide opportunities for students to build skills as environmental advocates and understand the vital need to protect our living world.

Through active, hands-on learning about the environment, children develop the knowledge and skills to address challenges in their communities while contributing to their own academic achievement.

Acknowledgments

Developed over five years of teaching outdoor environmental education, *Hop Into Action* is a collection of lessons for facilitating children's activities in the natural world. All creative works depend on experiences that preceded them, and this curriculum guide is no exception. In that light, I'd like to offer my thanks to those teachers, professors, naturalists, and other environmental education facilitators who have captured my attention and imagination and made my learning experiences both fun and meaningful. Finally, I owe my first debt of gratitude to my parents, who encouraged me to run free in the natural world, even if it meant coming home covered in both mud and duckweed and leaving a trail of smelly boots and other equipment to greet visitors at the door.

About the Author

David Alexander is an experienced naturalist who uses the environment to educate a diverse group of students at all age levels. He earned his graduate degree in environmental science, conservation biology at Green Mountain College and his bachelor's degree in natural resources at the University of Vermont. His enthusiasm and curiosity for the natural world is boundless.

Introduction

Hop Into Action is an amphibian education curriculum designed for grades kindergarten through fourth grade in a way that allows students to apply knowledge from one lesson to others in the field and classroom. This cross-disciplinary curriculum guide introduces children to the joy of amphibians through investigations that involve scientific inquiry and knowledge building, while treating science as a process and not as memorization. In turn, these lessons bridge the gap between knowledge and action by promoting critical thinking, problem-solving skills, and collaboration to help students become advocates for the environment.

Lessons are offered sequentially, but they may be used out of sequence if students are learning at the grade level listed and educators are familiar with students' prior knowledge. Lessons are tailored to allow for extensions to multiple learning styles as needed for students who experience and process information differently. For example, kinesthetic learners will be provided the opportunity to have a concrete experience feeling a frog in the lesson about amphibian identification or using clay to demonstrate metamorphosis. Visual learners will benefit when participating in the creation of lily pad Venn diagrams or frog pond habitat webs that allow students to represent information spatially. Auditory learners will benefit from the discussion built into each lesson and specifically benefit from the lesson "Audible Amphibians," which offers the opportunity to hear the calls of frogs and toads. Finally, learners who enjoy and benefit most from reading and writing will love the stories provided as resources and thrive when participating in the lessons "Ribbiting Discoveries in the Lily Pad Paper" and "Seasonal Discoveries Journal."

Lessons also provide students with an understanding of career pathways as they act as biologists, herpetologists, ecologists, reporters, and park naturalists to investigate frog ponds.

The lessons included are designed to take advantage of and exercise children's natural curiosity about the environment using observation, photographs, games, and direct instruction. The curriculum includes reference materials such as field guides, websites, and storybooks that complement lessons and allow for study of species found in your own region. I hope you and your students learn to love, appreciate, and protect amphibians as a result of the fun and educational ideas provided in this guide.

How to Use This Book

Each lesson plan includes basic information for the instructor to determine if the activity will meet his or her needs. The following information is provided: grade/ability level, subject area, skills used, class setting, time required, and group size. Lessons describe in detail the objectives or observable student outcomes of each lesson; method of meeting the objectives for each lesson; materials required to perform the lesson; background information that will help educate the instructor about the lesson topic; a procedure to follow for presenting each lesson; evaluation questions and methods to assess the knowledge of students after the lesson; extensions that provide additional study related to the lesson; and resource information that includes books, audiovisual references, and web resources.

The activities in this guide were designed to meet content standards outlined in the *National Science Education Standards* and the North American Association for Environmental Education's *Excellence in Environmental Education: Guidelines for Learning (PreK–12)*.

Educators should modify their use of lessons to meet the learning goals of their students' ages and ability levels. Permission is granted in advance for reproduction for purpose of classroom or workshop instruction. To request permission for other uses, send specific requests to publisher.

The following table outlines the grade levels and subject areas covered for each lesson:

Lesson Name	Grades	Subject Areas
How to Identify an Amphibian	K–4	Science, Language Arts, Art
Amphibian Encounter	K–3	Science, Language Arts
Amphibian Metamorphosis	K–2	Science, Drama, Art
Lily Pad Venn Diagrams	3–4	Science, Language Arts, Math
Frog Hop Relay Race	K–2	Science, Physical Education
Camouflaged Critters	K–2	Science, Art
Amazing Amphibian Migration	2–4	Science, Language Arts, Physical Education
Frog Pond Soup	3–4	Science, Language Arts
Frog Pond Web	3–4	Science, Language Arts

Lesson Name	Grades	Subject Areas
Frog Pond Lifeguard	4	Science, Language Arts
Audible Amphibian	1–4	Science
Feeding Frenzy	K–4	Science, Physical Education, Math
Salamander Smell	2–4	Science
Frog Pond Poetry	3–4	Science, Language Arts
Ribbiting Discoveries in the Lily Pad Paper	3–4	Science, Language Arts, Media, Art
Seasonal Discoveries Journal	3–4	Science, Language Arts
Herp, Herp, Hooray	4	Science, Language Arts
Frog Pond Choices	4	Science, Language Arts
Frogville Town Meeting	4	Science, Language Arts, Civics
Amphibian Art	2–4	Science, Art, Language Arts, History

Resource Information

National Research Council. 1996. *National science education standards.* Washington, DC: National Academies Press.

North American Association for Environmental Education (NAAEE). 2009. *Excellence in environmental education: Guidelines for Learning (preK-12).* Washington, DC: NAAEE.

Amphibian Curriculum Guide

Lesson Correlations to National Science Education Content Standards, Grades K–4

Content Standard	Topic	1	2	3	4	5	6	7	8	9	10	11	12	13	14	15	16	17	18	19	20
A. Science as Inquiry	Abilities necessary to do scientific inquiry		•	•			•			•	•									•	
	Understanding about scientific inquiry		•	•			•			•	•	•						•		•	
B. Physical Science	Properties of objects and materials								•												
	Position and motion of objects					•															
C. Life Science	The characteristics of organisms	•	•	•	•	•	•			•			•	•	•	•	•			•	•
	Life cycles of organisms			•						•										•	
	Organisms and environments		•		•	•	•	•		•		•	•	•	•	•	•		•	•	•
D. Earth and Space Science	Properties of Earth materials								•												
	Changes in the Earth and sky										•					•	•				

Amphibian Curriculum Guide

Lesson Correlations to National Science Education Content Standards, Grades K–4 *(cont.)*

Lesson

Content Standard	Topic	1	2	3	4	5	6	7	8	9	10	11	12	13	14	15	16	17	18	19	20
E. Science and Technology	Abilities of technological design							•			•							•	•	•	
	Understanding about science and technology										•	•						•	•	•	
	Abilities to distinguish between natural objects and objects made by humans								•	•											
F. Science in Personal and Social Perspectives	Characteristics and changes in populations							•		•										•	•
	Changes in environments							•		•										•	•
	Science and technology in local challenges																	•	•	•	
G. History and Nature of Science	Science as a human endeavor	•	•				•				•	•						•	•	•	•

NATIONAL SCIENCE TEACHERS ASSOCIATION

XVIII

Educating With Amphibians in the Classroom and Field

Amphibians in the Classroom

Due to the care and commitment required, educators should always receive permission from administrators before moving forward with housing amphibians in a nature center, recreation center, or classroom. Teachers should check with their board of education, school administrators, and the school nurse before housing amphibians in a classroom or handling them in the outside environment. Educators should have a plan in writing that describes any necessary funding and a care schedule that takes into account weekends and school breaks.

In a classroom, it is important that amphibians act as "Animal Ambassadors" that help educate students about their respective species. If, as an educator, you choose to host an animal ambassador, you may want to consult your students before making all the decisions. Consider the following questions:

- How would you create a habitat for the amphibian that provides basic needs and closely resembles the natural habitat? What does the amphibian need to survive? How can we create a habitat in our classroom that includes all of these things? Consider moisture, light, temperature, and food.
- What do amphibians need to eat? Do they eat the same things at all life stages?
- Should we hold and handle the amphibian?
- How long should we keep the animal in this artificial or model habitat?

Giving children the chance to help with this planning process will allow them to think deeply about the ways habitats meet the needs of animals. Ultimately, an amphibian in the classroom should be treated as an ambassador of its species and cared for with the utmost respect through responsible handling, feeding, maintenance, and general care schedules. Only one species of amphibian should be maintained in a classroom at one time, and it should not come into contact with anything else shared with other animals unless sterilized for health and safety reasons.

There are many biological science suppliers that offer live amphibians for classroom use; however, they may only be available at certain life stages throughout the seasons and require planning ahead. *After the completion of the curriculum, amphibians purchased for use in the classroom should not be released into nearby habitats, but rather kept and cared for until the end of their lives.* You may find your local herpetological society helpful at finding someone who can provide the care needed or at last resort a veterinarian can euthanize according to the American Veterinary Medical Association's (AVMA) Guidelines on Euthanasia.

This is important because releasing captive amphibians can spread disease or organisms against which native wildlife may not have immunity. The released amphibians may also not be native to the habitat or not have time to adjust to the seasonal changes taking place and therefore not capable of survival.

If amphibians are collected outside for short-term classroom use (see permits on following page) they may be released at the capture site so long as proper sanitation procedures were followed (as outlined on the following page).

There are many opportunities to educate students about how to care for amphibians. Anne Mazer's book *The Salamander Room* is a great place to start a discussion regarding the responsibilities involved for younger learners. It is also important to distinguish the fantasy from reality found in books and movies so students begin thinking about the basic needs of living things and how they are met.

Amphibians in the Field

The best place to learn about the environment is the natural environment. The proximity of a pond to a classroom allows a much more intimate relationship with nature in terms of students being able to observe it with relatively little time and effort involved. It also helps to use local natural and cultural surroundings as the context for instruction and learning. When this place-based education is implemented, students and community members can benefit from partnerships.

Many students will benefit from opportunities to move in and out of open and focused explorations in a natural setting. When students are asked to focus their attention toward work, reading, or tests, they can feel fatigued. When they have opportunities for open exploration, involuntary attention can take over, giving the brain time to relax, in turn leading to better behavior and concentration. In addition, offering students playful learning opportunities can lead to better academic success and both interest in and excitement about the subject area.

Students should be briefed about what they might see outside, as some may be timid or scared if they have not had experience exploring the natural world. You should also discuss general discipline with students, including school rules that apply while outside.

Amphibian habitats may be found with assistance from a local environmental center, parks and recreation department, or state fish and wildlife agency. If your organization does not have access to a pond or other suitable habitat, you should consider creating a backyard or schoolyard habitat. Excellent resources exist from both the National Wildlife Federation and Tree Walkers International that will help you build a pond habitat suitable for amphibians. Remember that depending on the distance and weather conditions, field outings may require field essentials such as rain gear, rubber boots, waders, waterproof notebooks, and more (review Safety Practices for Outdoors and in the Classroom, p. xxiii). For younger students, you may even choose to consider life jackets as a safety precaution.

Handling Techniques

As with handling all life, we must show our students how to be respectful. It is important that all handlers wash their hands before and after holding or touching an amphibian. Improper handling of amphibians can be detrimental to their health, so an adult should always be present to assist. One technique that should be encouraged is to sit low while holding an animal so if it squirms or hops there won't be an injuring fall. Gently touching with one finger should also be encouraged.

Things to remember:
- Return amphibians to the same location where they were found. If found under a log or rock, place the amphibian next to the cover item and return the cover item as it was found. Consider the cover item to be similar to the roof of a house. It maintains a microclimate that the amphibian requires.
- Avoid getting insect repellent, sunscreen, or other personal care products on hands, as it may absorb into the skin of the amphibian.
- If you are exploring outside environments, be careful not to disturb the habitat you wish to study. You may remind younger students of this by explaining that "plants grow by the inch but die by the foot."

Disinfectant Techniques

Before and after placing amphibians in tanks or in contact with equipment—including nets, filters, and boots—a disinfectant should be applied to the equipment. First clean with a detergent and rinse clean prior to bleaching. A 1% solution of household bleach (usually a 4% solution of sodium hypochlorite) can be made using one part household bleach to three parts water; a minimum contact time with equipment of 15 minutes is necessary.

It is also important to age chlorinated water for 24 hours, or use a drop of Chlor Out to dechlorinate water before introducing the animal, or the chlorine can harm it.

Permits

Care should be taken to acquire all necessary information and permits before purchasing or collecting wildlife, as some species may be threatened or endangered. There may be national laws as well as state laws that restrict and regulate what species are available for outside handling or inside education. Your state department of environmental protection or fish and game should be able to provide you with a list of protected species and permit applications.

Resource Information

American Veterinary Medical Association (AVMA). *www.avma.org*

Mazer, a. 1994. *The salamander room*. New York: Dragonfly Books.

Mendelson, J. 2009. Considerations and recommendations for raising live amphibians in classrooms. Society for the Study of Amphibians and Reptiles, *www.ssarherps.org/documents/amphibians_in_classroom.pdf*.

National Research Council. 1996. *National science education standards*. Washington, DC: National Academies Press.

National Science Teachers Association. Responsible use of live animals and dissection in the science classroom. NSTA. *www.nsta.org/about/positions*

North American Association for Environmental Education (NAAEE). 2009. *Excellence in environmental education: Guidelines for Learning (preK-12)*. Washington, DC: NAAEE.

Wyzga, M. 1998. *Homes for wildlife: A planning guide for habitat enhancement on school grounds*. Concord, NH: New Hampshire Fish and Game Department.

Safety Practices for Outdoors and in the Classroom

Outdoors

1. Teachers should always visit outdoor areas to review potential safety hazards prior to students carrying out activities.
2. Keep clear of outdoor areas that may have been treated with pesticides, fungicides, or other hazardous chemicals.
3. When working outdoors, students should use appropriate protective equipment, including safety glasses or safety goggles (if working with hazardous chemicals), gloves, closed-toed shoes, long-sleeve shirts and pants.
4. Caution students of poisonous plants (e.g., poison ivy, sumac), insects (e.g., bees, ticks, mosquitoes), and hazardous debris (e.g., broken glass).
5. Teachers need to inform parents in writing of on-site field trips relative to potential hazards and safety precautions taken.
6. Teachers need to check with the school nurse relative to student medical issues (e.g., allergies, asthma). Be prepared for medical emergencies.
7. Teachers need to have a means of communication (cell phone, two-way radio) in case of emergencies.
8. Wash hands with soap and water after doing activities outdoors.
9. Contact the main office prior to taking classes out of the building.

In the Classroom

1. Always review Material Safety Data Sheets (MSDS) with students to go over safety precautions in working with hazardous chemicals.
2. Remind students only to observe animals. Do not touch or pick up animals unless instructed to do so by the teacher.
3. Use caution in working with sharp objects such as scissors or glass slides.
4. Wear protective gloves when handling animals.
5. Do not eat or drink anything in the science lab or when handling animals.
6. Wash hands with soap and water after doing activities with hazardous chemicals, soil, or biologicals (plants or animals).
7. To disinfect cages and other equipment used in animal care, first wash the items in hot water with detergent. Scrape off stuck material. Rinse with plain water. Apply a bleach solution (½ cup household bleach to 1 gallon of water). Let cage and equipment sit in the bleach solution for a minimum of 20 minutes. Rinse again with plain water.
8. Use caution when working with clay. Dry or powdered clay contains a hazardous substance called silica. Only work with and clean up clay when wet.

Lessons for Prekindergarten Learners

While this guide was designed to provide comprehensive lessons to educators in kindergarten through fourth grade, provided here are lesson ideas for prekindergarten educators looking to add live event learning and hands-on science for little explorers.

Slime: Have children experience an amphibian-like substance with their sense of touch. Place the following ingredients in a plastic bag: 1 cup cornstarch, ½ cup water, green food coloring. Remove the air and knead the bag slowly until the mixture is well blended. Allow children to touch and play with the slime, but remind them not to taste it!

Getting Dirty: Have children play and experiment with mud (know the source of the dirt and make sure it is free of pesticides). Wash hands with soap and water after the lesson.

Sink or Float: Have children experiment with a water table to determine what sinks or floats. Predictions followed by results can be recorded on a chart.

Follow the Frog: Place amphibian pictures along a trail for children to find on their walk to the pond.

Hop Around: Follow the leader and imitate the movements of amphibians or other pond life.

Matching Frogs: Create a matching game where children must find like colors of frogs and pair them together. Or pair baby and mommy animal pictures.

Feeding Frenzy: Have children search around a field for "frog food" or strips of felt that represent the different foods frogs would find. Discuss what colors were easiest to find and why.

Tools of the Trade: Have children practice using hand lenses to discover details and see more in nature.

Seasonal Discovery: Have children revisit a pond or natural area monthly so they have opportunities to notice as much seasonal change as possible. Scavenger hunts can be added to the nature walk.

Frog Puppets: Have children paste premade pieces onto a paper bag to create their own frog puppets.

Popsicle Puppets: Have children color animals and paste them on sticks. They can be raised and lowered when the animals appear in a story or cast onto a sheet as shadows.

Prekindergarten Books

Faulkner, K. 1996. *Wide mouthed frog*. New York: Dial

Fleming, D. 2007. *In the small, small pond*. New York: Henry Holt.

Kent, J. 1982. *The caterpillar and the pollywog*. New York: Aladdin.

Lescroart, J. 2008. *Icky sticky frog*. Atlanta: Piggy Toes Press.

Lionni, L. 1996. *It's mine*. New York: Dragonfly Books.

Pallotta, J., and R. Masiello. 1990. *The frog alphabet book*. Watertown, MA: Charlesbridge.

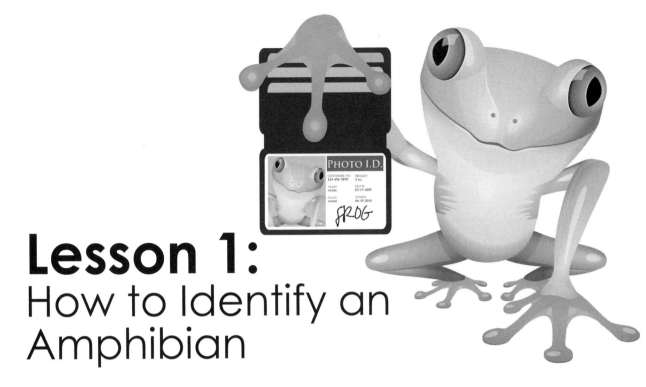

Lesson 1:
How to Identify an Amphibian

Objectives

Students will be able to identify the physical characteristics of an amphibian and explain that amphibians come in a variety of forms, colors, and adaptations.

Method

Students in grades K, 1, and 2 will observe and discuss the characteristics of an amphibian.
Students in grades 3 and 4 will also collect and analyze data based on their observations.

Materials

Display board, amphibian structure worksheet, amphibian identification worksheet, clipboard, pencil, photographs of amphibians

Grade Level: K–4

Subject Area: science, language arts, art

Skills: description, identification, drawing, small-group work

Setting: Inside and outside

Lesson Duration: 30 minutes

Group Size: no minimum size

National Science Education Standards (Grades K–4):
- **Life Science:** The characteristics of organisms
- **History and Nature of Science:** Science as a human endeavor

Background Information

Amphibians are animals that generally live both on land and in water. Amphibians first appeared in the Devonian Period hundreds of millions of years ago and were around before, during, and after the time of the dinosaurs (Duellman and Trueb 1986). Evolutionary amphibians are the first tetrapods (four-legged animals) and were ancestors to the dinosaurs. Amphibians have adapted to survive around the world in a variety of habitats, including forests, fields, wetlands, prairies, deserts, and even your own backyard. People who study amphibians are called herpetologists.

Amphibians are cold-blooded creatures and therefore must use the environment to help regulate their body temperatures. Most amphibians spend part of their lives in water and part on land. They hatch from eggs and change as they grow.

Most adult frogs, salamanders, and caecilians share similar physical characteristics, such as soft, moist, slimy, permeable skin. However, they also can be differentiated by their physical characteristics to determine if they are a frog, salamander, or caecilian. With careful observation of photographs, drawings, or live amphibians, students will be able to categorize frogs as having webbed feet and a round tailless body; students will recognize salamanders as having a longer body than a frog and a tail; and students will note that caecilians have grooves that form rings around the body and no legs.

Amphibians that make up each order may appear similar at first glance but with practice students will begin to describe and differentiate between their identifiable characteristics—such as shape, color, and pattern—to determine their specific species.

Procedure

1. Show students photographs of amphibians using books and other resources listed at the end of this lesson. Ask students to point out the similarities and differences between our bodies and theirs. For example, we share many of the physical characteristics of frogs and salamanders, including eyes, nose, mouth, feet, head, and legs.

2. Draw an amphibian on the board and ask students to come up and name the parts they recognize. Provide students in grades 2 through 4 with the Amphibian Structure worksheet and ask them to write the descriptive words that match the body parts after they are discussed and written out on the display board. Prewriters can color in the body parts of each amphibian as they are labeled on the board. The parts of an adult frog that can be labeled include head, webbed feet, ears, mouth, nose, eyes, and body. The parts of an adult salamander that can be labeled include mouth, nose, eyes, tail, toes, head, front legs, hind legs, and body. The parts of a caecilian include eyes, mouth, and grooves that form rings around the body.

3. Discuss how the amphibian's body parts help it function. For example, some have webbed feet to swim, ears to hear, and a mouth to eat.

4. Discuss with students the importance of amphibians. Answers may include that they are fun to catch and observe, they eat bugs, and they are eaten by other animals for food.

5. If an outside environment is available, discuss expectations and outdoor safety inside before you take students in search of live amphibians in their habitat. Instruct students in grades 2 through 4 to act as herpetologists

and follow the instructions on the Amphibian Identification worksheet to observe and identify an amphibian if they are found outside. Students may complete all or part of the worksheet depending on ability level. Encourage students to move slowly and not scare the amphibian so it can be recorded in their sketches and notes. They should take time to observe the amphibian's unique body parts and think about how they function to help the animal survive. If no outside environment is available, you may choose to have a captive amphibian available for study in the classroom.

6. Allow students to use age-appropriate field guides to identify findings based on their memory and recorded description. *Peterson First Guide to Reptiles and Amphibians* will have drawings and descriptions of amphibians found in a North American pond.

Reflect and Explain

*   What are some things you've observed about amphibians?
*   What are some ways amphibians and humans are similar or different?
*   How do amphibians differ in size, shape, body parts, and behaviors? How are they alike and different?

Extensions

*   Play a game of charades using "pond etiquette" as the theme. Ethical guidelines may include that if you turn over a rock, you should put it back; touch with one finger; let nature do the talking; and wash hands and tools after contact with a habitat or organism.
*   Take a trip to a zoo or aquarium to see captive amphibians, or visit a nature center or park to observe living amphibians in their natural habitats.
*   Begin journaling your experiences with amphibians from the first to the last experience!

Resource Information

Bakken, A. 2006. *Uncover a frog.* Berkeley, CA: Silver Dolphin Books.

Clarke, B., and L. Buller. 2005. *Amphibian.* New York: DK Children.

Conant, R., R. Stebbins, and J. Collins. 1991. *Peterson first guide to reptiles and amphibians.* Orlando, FL: Houghton Mifflin Harcourt.

Duellman, W. E., and L. Trueb. 1994. *Biology of amphibians.* Baltimore, MD: The Johns Hopkins University Press.

Kalman, B. 2000. *What is an amphibian?* New York: Crabtree Publishing.

Mattison, C. 2007. *300 frogs: A visual reference to frogs and toads from around the world.* Ontario, Canada: Firefly Books.

Sill, C. 2001. *About amphibians: A guide for children.* Atlanta, GA: Peachtree.

Amphibian Structure

Name(s):_____

Label the Parts of the Amphibian

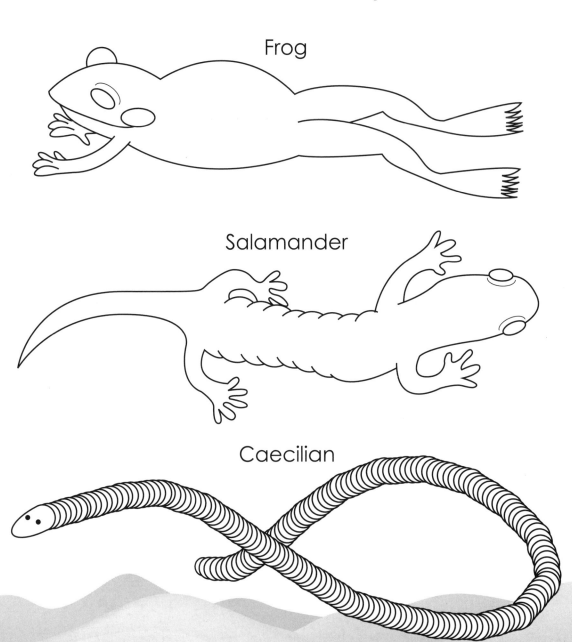

Frog

Salamander

Caecilian

NATIONAL SCIENCE TEACHERS ASSOCIATION

Amphibian Identification

Name(s): _____   Date: _____

Weather: _____   Location: _____

Circle the matching descriptions.

Frog Size: Ping-Pong ball   Tennis ball   Softball   Basketball

Salamander Size: Half a pencil   Marker   Stapler   Water bottle

Habitat: In water   On land   In a tree or shrub   Other

Color: Green   Brown   Blue   Orange   Gray   Black

Pattern: Spots   Stripes   Warts

Ridges on the Back? Yes   No

Describe or illustrate interesting characteristics about the amphibian you're studying.

Do you know what species you are observing?

Behavior

Describe what the amphibian is doing.

If the amphibian is calling, describe the sound it's making.

Lesson 2:
Amphibian Encounter

Objectives
Students will be able to identify the behaviors of amphibians in their natural habitat.

Method
Students explore an amphibian habitat to observe and collect data about amphibian behaviors.

Materials
Amphibian Activity Search worksheet, pencil, pen

Background Information
To be successful in a career involving animals, you must spend time carefully observing them. Herpetologists study amphibians and reptiles. Amphibians are found on all continents except Antarctica. They are easiest to find in spring, summer, and early fall but difficult to find in winter, as they will dig themselves underground, hide in an animal hole, or even rest in the mud at the bottom of a pond to wait out the winter.

When amphibians are found, they may be hiding, hopping, becoming food, laying eggs, calling, or basking. They are often easiest to find near sources of water, as they must keep their skin moist to assist in breathing, and most also

Grade Level: K–3

Subject Area: science, language arts

Skills: description, identification, small-group work

Setting: outside and inside

Lesson Duration: 30 minutes

Group Size: no minimum

National Science Education Standards, Grades K–4:
- **Science as Inquiry:** Abilities necessary to do scientific inquiry
- **Science as Inquiry:** Understanding about scientific inquiry
- **Life Science:** The characteristics of organisms
- **Life Science:** Organisms and environments
- **History and Nature of Science:** Science as a human endeavor

require water as a place to lay their eggs. You must look slowly and closely to discover amphibians because they may be as small as half an inch and camouflaged. Adult frogs are also likely to leap into the safety of the water if you approach too quickly. If you do not have an available habitat to find amphibians, re-evaluate your habitat with help from Tree Walkers International or the National Wildlife Foundation.

Procedure

1. Inform students they will be going on a field walk acting as herpetologists in search of amphibians. (Review Safety Practices for Outdoors, p. xxiii.)

2. Ask students how they think herpetologists behave while observing amphibians in the field. List and discuss what they say. Ask next what behaviors they think amphibians will be displaying. List and discuss, guiding them toward the desired answers. Finally, discuss what they should do to have the best chance of discovering amphibians. List and discuss; examples may include listening and walking slowly while searching the habitat or turning over rocks and logs. Remember to inform students that they should return all rocks and logs as they were to avoid disrupting the habitat.

3. Tell students they will act as herpetologists and take along the Amphibian Activity Search worksheet to record their observations. When an amphibian is found, its behavior (hiding, hopping, becoming food, laying eggs, calling, basking) will be recorded on the amphibian behavior worksheet. Let students know that at times the behavior may be undetermined, so they may choose the behavior they think best matches the action observed.

4. If there is no outside environment to explore, place pictures of the amphibians on the Amphibian Activity Search worksheet along a hallway or around an indoor pretend pond environment for students to discover.

Reflect and Explain

- What behaviors or activities helped you as the herpetologist find amphibians?
- What amphibian behaviors did you observe? Why do you think the amphibians were behaving that way as you observed?

Extensions

- Discuss what a herpetologist would carry and follow up the discussion by dressing up a student with a net, boots, a field guide, and a notebook.
- Have students research specific amphibians and include information

on life cycle, habitat, and diet so final work can be used to create a field guide, calendar, or poster that can be shared with the community. Funds can be raised for an environmental cause if the compiled project is sold.

- Report findings to Frog Watch USA: *www.aza.org/frogwatch.*
- Provide collected data to the local conservation commission, town planners, state fish and wildlife service, and other interested organizations.
- If you find malformed frogs, contact your local fish and wildlife service.
- Discuss how amphibians are interconnected with the environment using the frog pond web lesson.

Resource Information

Conant, R., and J. T. Collins. 1998. *Peterson field guide to reptiles and amphibians, Eastern and Central North America.* 3rd ed. Boston: Houghton Mifflin.

Fish and Wildlife Service Offices, State, Territorial, and Tribal. *www.fws.gov/offices/statelinks.html.*

Hawes, J. 2000. Why frogs are wet. New York: Harper Collins.

National Wildlife Federation Garden for Wildlife. *www.nwf.org/gardenforwildlife.*

Tree Walkers International. Operation Frog Pond. *www.treewalkers.com.*

Amphibian
Activity Search

Name(s):_____

Date: _____

Weather:_____

Place a check mark next to the amphibian behavior when observed.

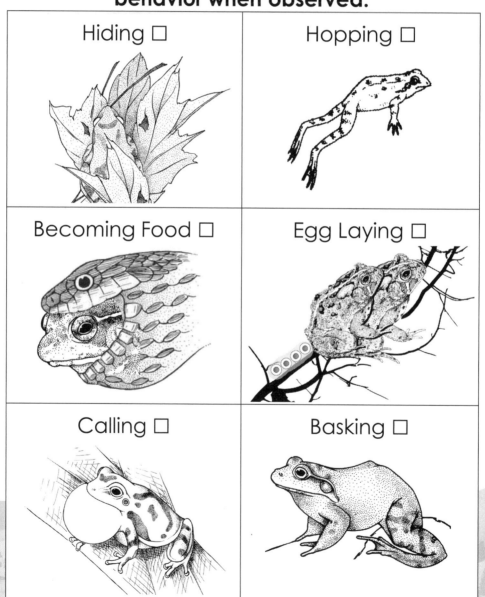

Hiding ☐

Hopping ☐

Becoming Food ☐

Egg Laying ☐

Calling ☐

Basking ☐

Lesson 3:
Amphibian Metamorphosis

Objectives
Students will be able to identify the life stages of amphibians from egg to adult.

Method
Students learn the life cycle of a frog from egg to adult through discussing and acting out their growing stages.

Materials
Amphibian Metamorphosis worksheets, pencil or colored pencils, pictures of amphibian eggs, larva or tadpole, and adult frogs

Background Information
Amphibians go through a changing process known as **metamorphosis** as they grow from an **egg** to **larva** (tadpole) to **adult** (frog). They are the only **vertebrates** to undergo

Grade Level: K–2

Subject Area: science, drama, art

Skill: description

Setting: inside or outside

Lesson Duration: 30 minutes

Group Size: no minimum size

National Science Education Standards, Grades K–4:
- **Science as Inquiry:** Abilities necessary to do scientific inquiry
- **Science as Inquiry:** Understanding about scientific inquiry
- **Life Science:** The characteristics of organisms
- **Life Science:** Life cycles of organisms

this transformation. Metamorphosis can be split into two parts: *meta*, meaning change, and *morph*, meaning shape. Different amphibians go through this process in different ways. Eggs of most frogs are fertilized externally in the water as the male holds onto the female in a grasp known as **amplexus**. The water will act as the amphibian's nursery or incubator at this stage. The frog eggs will then metamorphose into legless tadpoles with feathery **gills** that enable them to breathe dissolved oxygen in the water. Next the tadpoles will grow hind legs, followed by front legs and begin to transition into an adult breathing with **lungs** that can live outside the water.

Salamanders, in contrast, are fertilized internally. Many species, particularly those that require **vernal pools** to lay eggs, will take part in a night dance where the male will drop a **spermatophore,** or sperm packet that is picked up by a female with her cloaca for internal fertilization before eggs are laid.

The amount of time a frog tadpole or salamander larvae takes to become an adult varies depending primarily on the **species** but also may be affected by water temperature and availability of food. Most amphibian young never reach adulthood as they become prey to **aquatic** and **terrestrial** predators hunting from above and below. For the best chance at survival, amphibians have adapted a variety of techniques.

Green frogs will lay eggs in a thin film of jelly floating on the surface of a pond or lake that provides water for an extended period of time and allows for the full development of the frog before the pool dries up. The eggs hatch into tadpoles with strong tails for swimming, scraper mouths, and gills. The tadpoles also often have big stomachs to hold lots of food. As the season progresses, the tadpole's tail is absorbed into the body to provide nourishment for the transition ahead, hind legs begin to grow (followed by front legs), and gills disappear and lungs develop, allowing the frog to leap onto the land and breathe, so long as their skin remains moist.

The red-backed salamander does not require a body of water to lay its eggs. The female will attach her eggs to the ceiling of a rock or log and stay with the eggs until they hatch as fully formed salamanders, ready to go off and forage on their own. Another amphibian species, the red-spotted newt, carefully deposits each egg on its own, often with a leaf wrapped around it to conceal the young until they are able to swim off on their own.

In addition to these development strategies, many species of amphibians seem to believe "there is no place like home" and have site fidelity, which means they will return to lay their eggs in the same general location as they were born. If the pool or the habitat on the way to the pool was disturbed, these amphibians would have difficulty reproducing because they would not find their egg-laying location. Observers have even reported finding eggs on blacktop or construction fill after breeding pools were covered.

It's important that amphibians have healthy habitats in which to breed because they are particularly susceptible to environmental pollutants. These animals are often called **indicator** species, because as they transition or metamorphose, artificial hormones or **pollutants** in the amphibian's habitat can interfere with the growth process, and deformity or even death can occur.

Procedure

Part 1

1. Play is a fundamental part of children's development, so as the facilitator of this activity, you can engage the students in the act of using their bodies to represent the changing life stages of an amphibian.
2. Prior to demonstrating the life stages of an amphibian, discuss the life stages of a human being, including infant, toddler, child, teenager, adult, and senior. Explain that we can recognize ourselves in baby pictures, but amphibians don't look like small versions of themselves.
3. Instruct the students to follow the movements you make and repeat the name of the life stage that is being imitated. Word cards can also be created for students to see the vocabulary they are learning.

Frog Metamorphosis

4. Egg: Wrap arms around body and crouch low.
5. Tadpole: Slowly wiggle body, shake your bottom to represent the tadpole tail and breathe with gills as demonstrated by making a fish face (squeezing cheeks into mouth).
6. Pollywog: Slowly release your arms from around your body and make swimming strokes with arms pushing water away from face.
7. Adult frog: Hop around the classroom as an adult frog.

Salamander Metamorphosis

8. Egg: Wrap arms around body and crouch low.
9. Larvae or metamorph: Slowly wiggle, shake tail, and breathe with gills (fish face)
10. Adult salamander: Crawl around the classroom as an adult salamander.

Part 2

Preplanning

11. Create a station where students observe one or more stages of frog metamorphosis. Remind students only to observe and not to touch anything. The station may be set up either in the classroom or out in the field at a frog pond study site. Refer to Educating With Amphibians in the Classroom and Field (p. xix) as well as the Resource List (p. 111) for reminders on disinfection, collection, and safety.
12. Include a tank for captive viewing and magnifying glasses so students may observe the life stage in detail. If you begin with eggs or larvae, return to the observation station to complete the worksheet as they metamorphose into adults.

Procedure

13. Provide the Amphibian Metamorphosis worksheets to complete in the field or in class at the observation station.
14. Instruct students to make observational drawings of the stages of the frog in the respective squares and write about their observations alongside the squares on the observation station worksheet.

Reflect and Explain

- Ask students to act out the movement of the different stages of amphibian growth without the leader guiding the movement.
- Evaluate student ability to complete Amphibian Metamorphosis worksheets.
- Draw and cut out an amphibian life cycle and have the students place the steps in sequence. You may have students tie each to a string on a coat hanger to make a wire mobile or paste the steps into a flip booklet.

Extensions

- Use modeling clay to create an amphibian that comes "alive." Start making an egg or ball that is reshaped to morph into the continuing stages until adulthood.
- Use bubble wrap to make replicas of what different frog, toad, salamander, or newt eggs look like in a pond.
    o Green frog: floating raft of eggs
    o American toad: long strands of eggs that look like beads on a necklace
    o Spring peeper: individually placed eggs on submerged vegetation
    o Spotted salamander: clump of eggs attached to submerged vegetation
- Set up an investigation by counting the number of days that each stage of metamorphosis takes until your amphibian eggs become adults. Create a class graph to post and discuss. Begin a discussion about the scientific method using the following discussion questions:
    o Will an amphibian go through metamorphosis faster if provided higher or lower temperature water?
    o Do all amphibians take equal time to go through metamorphosis? How can we find out?

Resource Information

Biological Supply Catalogs to Purchase Tadpoles

Carolina Biological Supply: *www.carolina.com*

BioQuip Products: *www.bioquip.com*.

Metamorphosis Storybooks

Godwin, S. 1999. *The trouble with tadpoles: A first look at the life cycle of a frog*.
London: Hodder Wayland.

Himmelman, J. 1998. *A salamander's life*. Danbury, CT: Children's Press.

Himmelman, J. 1998. *A wood frog's life. Danbury,* CT: Children's Press.

Kaufman, B. 2006. *The life cycle of a frog*. New York: Crabtree.

Pfeffer, W. 1994. *From tadpole to frog*. New York: Harper Collins.

Amphibian Metamorphosis

Name(s):_____

Date: _____

Draw and describe the amphibian's appearance at each stage of metamorphosis.

Egg

Tadpole

Adult

Amphibian Metamorphosis

Name(s): _____

Date: _____

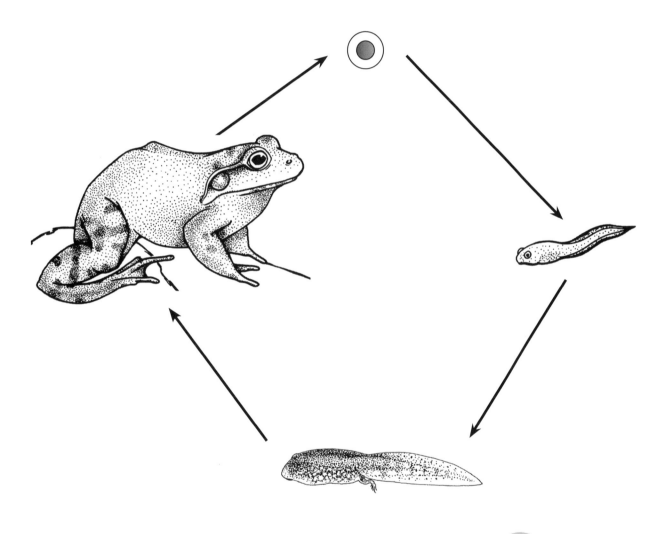

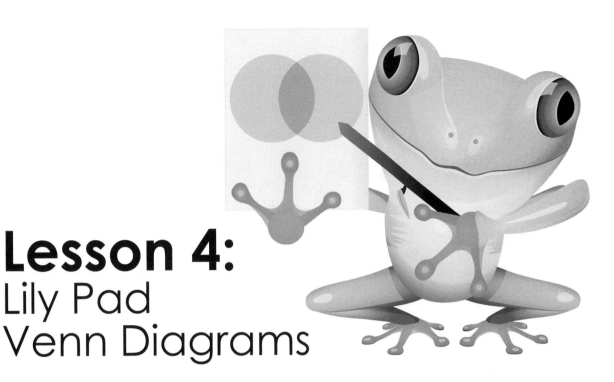

Lesson 4:
Lily Pad Venn Diagrams

Objectives

Students will be able to identify the similarities and differences between amphibians as they compare and contrast two different species.

Method

Students compare and contrast the similarities and differences between amphibians using Venn diagrams.

Materials

Lily Pad Venn Diagrams worksheets, pencil, visual reference to amphibians

Background Information

Grade Level: 3–4

Subject Area: science, language arts, math

Skills: analysis, description, identification

Setting: inside or outside

Lesson Duration: 30 minutes

Group Size: no minimum size

National Science Education Standards, Grades K–4
- **Life Science:** The characteristics of organisms

Amphibians come in many shapes, sizes, and colors and can be described and identified using language that appeals to our senses of sight, touch, and hearing. When students compare different amphibians, they are more likely to think deeply about the physical and behavioral characteristics that make each unique. Students may discover unique characteristics that showcase the differences between amphibians and also begin to recognize their similarities. An example of the physical differences students may observe is that adult salamanders and newts have a tail and adult frogs do not. Students may also visually notice the behavioral characteristics that are unique after watching frogs jump and salamanders walk, or they may listen and notice that frogs and toads can be told apart by their calls but salamanders and caecilians do not call. Students will also notice that most amphibians are cold-blooded, have moist skin, and undergo metamorphosis.

Procedure

1. Read a story or show pictures to begin a discussion about the physical and behavioral characteristics of amphibians.

2. Engage students in a descriptive conversation about the similarities and differences between frogs, toads, salamanders, newts, or caecilians. You may first choose to create a T-chart to help graphically organize student answers, listing characteristics of one amphibian on one side of the chart and characteristics of another on the other side of the chart. This can be started by asking students an open-ended question, such as what they notice about each animal or what the book says about each animal.

3. Introduce students to Venn diagrams by inserting their answers from the T-chart into the circles on the Venn diagram. Ask them what the amphibians have in common and fill in the overlapping portion that includes shared characteristics.

4. Depending on the ability of students, you may choose to provide a list of key characteristics to be used in the Venn diagram or allow students to research these characteristics on their own or in groups.

5. If available, show students the species side by side, with animals in a tank or outside, so that they may compare and contrast while observing the living species. If living species are not available, prepare pictures for visual reference.

Reflect and Explain

- What are the similarities and differences between the amphibian species you compared and contrasted?
- Based on similarities between amphibians, how would you define an amphibian?
- Attach a picture of one frog and one toad on each side of a display board or in a T-chart. Create text cards with characteristics that can be attached to the board underneath the animal or in between if the characteristic is shared by both the frog and toad.
- Place amphibian pictures into a pond net. Ask students to pull cards out of the net and decide if they are a frog, toad, salamander, newt, or caecilian. They should explain why.
- Write a comparison essay using the Venn diagram or T-chart that was created.

Extensions

- Create a Venn diagram to compare and contrast amphibians and humans.
- Create a Venn diagram to compare reptiles and amphibians. Discuss the slimy skin of amphibians and the scaly skin of reptiles.
- Make up a story about an amphibian or other animal using vocabulary that describes their characteristics. See if the group can figure out what amphibian the student was discussing based on descriptive writing.

Resource Information

Kalman, B. 2000. *What is an amphibian?* New York: Crabtree.

Zim, H., H. Smith, and J. Gordon. 2001. *Reptiles and amphibians: Revised and updated.* New York: St. Martin's Press.

Frogs

Bulging eyes

Jump

Have vocal sacs

Fertilized externally

Both

Are amphibians

Hatch from eggs

Young have gills

Young live in water

Adults have lungs

Have a backbone

Have vocal sacs

Go through complete metamorphosis

Toads

Dry skin

Poison glands behind eyes

Spend most time on land

Have no teeth

Lay eggs in chains

Walk

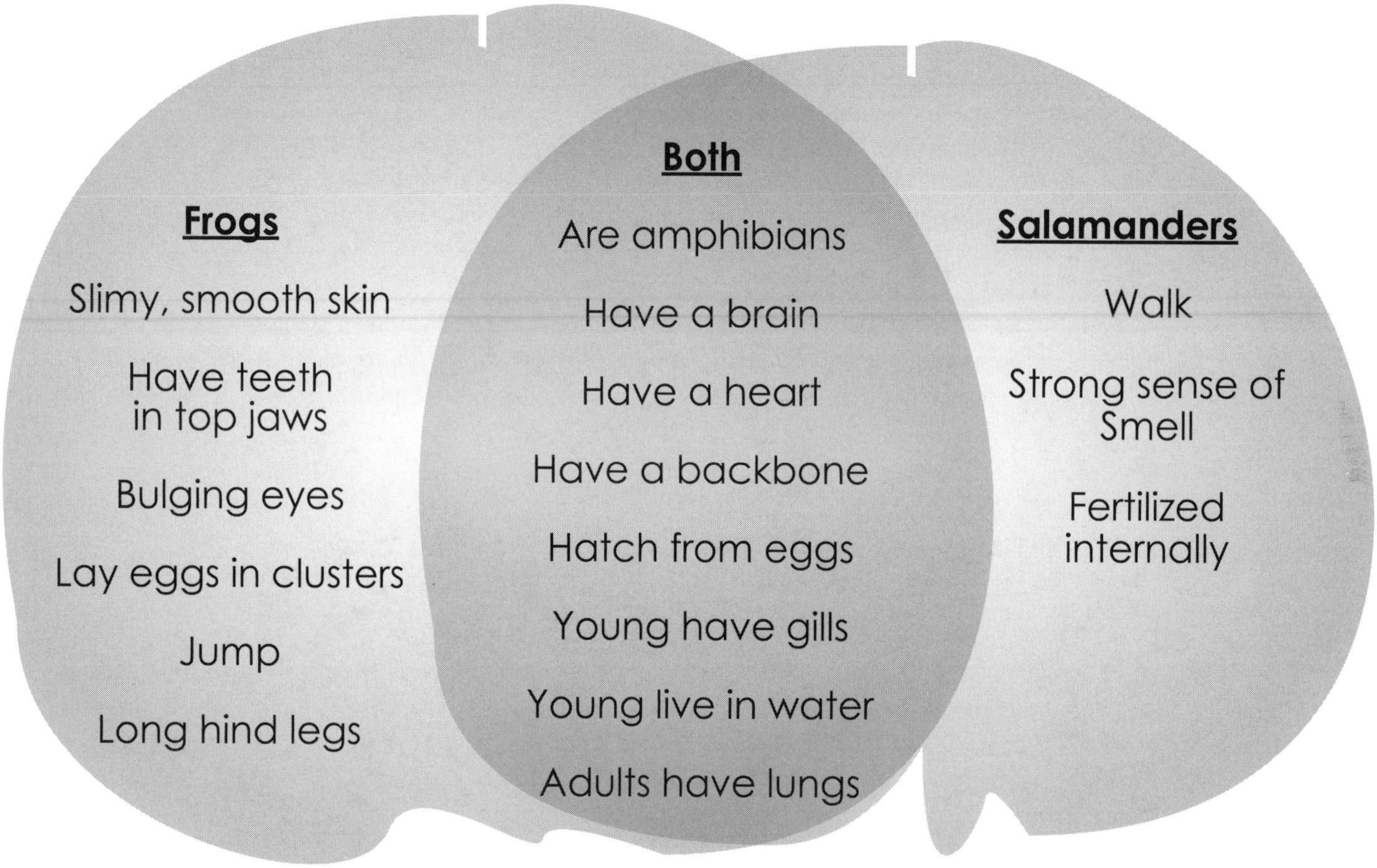

Frogs

Slimy, smooth skin

Have teeth
in top jaws

Bulging eyes

Lay eggs in clusters

Jump

Long hind legs

Both

Are amphibians

Have a brain

Have a heart

Have a backbone

Hatch from eggs

Young have gills

Young live in water

Adults have lungs

Salamanders

Walk

Strong sense of
Smell

Fertilized
internally

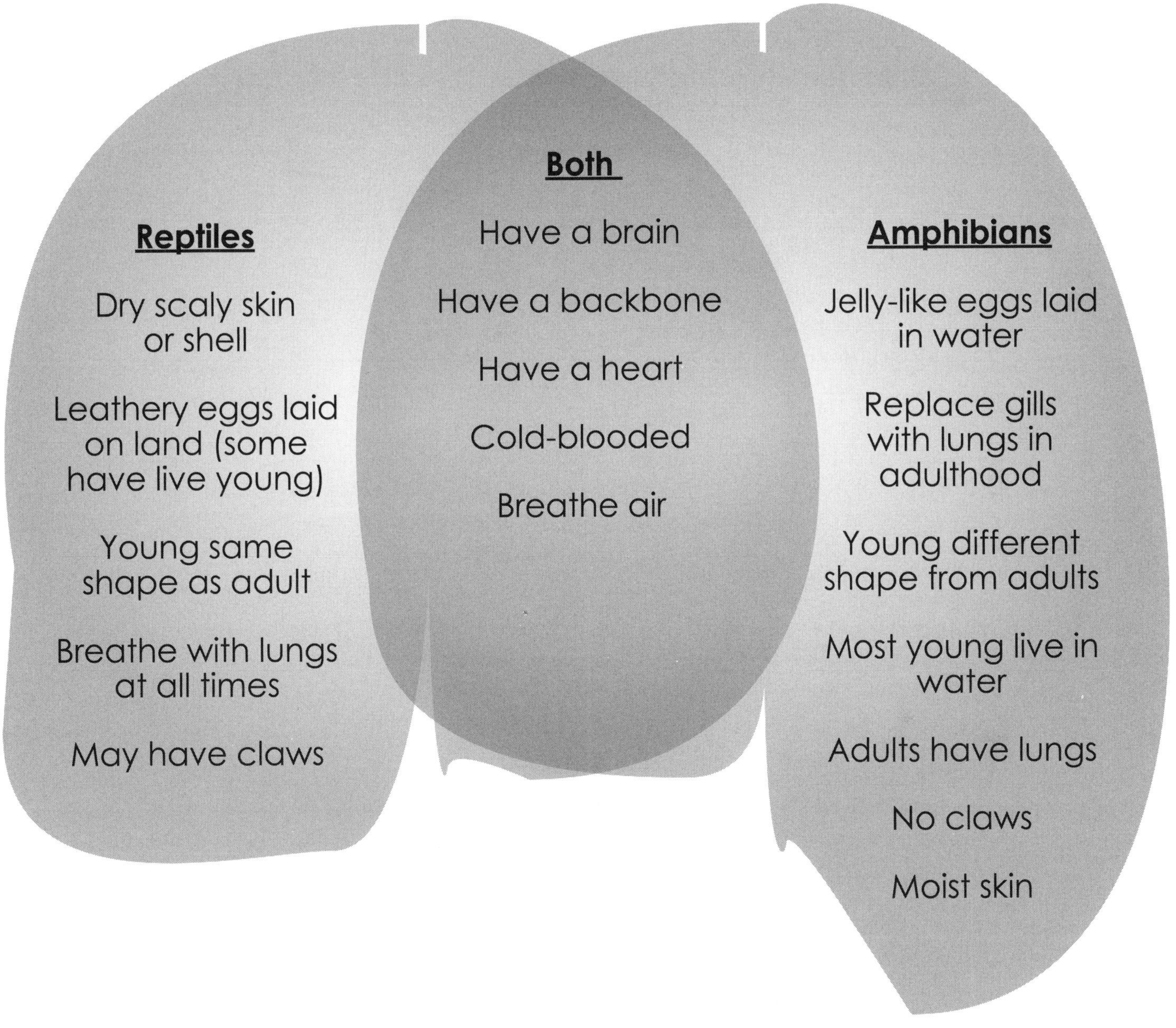

Both

Have a brain

Have a backbone

Have a heart

Cold-blooded

Breathe air

Reptiles

Dry scaly skin
or shell

Leathery eggs laid
on land (some
have live young)

Young same
shape as adult

Breathe with lungs
at all times

May have claws

Amphibians

Jelly-like eggs laid
in water

Replace gills
with lungs in
adulthood

Young different
shape from adults

Most young live in
water

Adults have lungs

No claws

Moist skin

Lily Pad Venn Diagrams

Name(s):_____

Write the descriptive characteristics into the frogs and toads Venn diagram.

- Are amphibians
- Jump
- Walk
- Hatch from eggs
- Young live in water
- Adults have lungs
- Have a backbone
- Have vocal sacs
- Go through complete metamorphosis
- Spend most time on land

- Bulging eyes
- Lay eggs in clusters
- Lay eggs in chains
- Young have gills
- Long hind legs
- Slimy, smooth skin
- Dry, watery skin
- Poison glands behind eyes
- Have no teeth
- Have teeth in top jaws

Both

Frogs

Toads

Lily Pad Venn Diagrams

Name(s): _____

Write the descriptive characteristics into the frogs and salamanders Venn diagram.

- Bulging eyes
- Jump
- Strong sense of smell
- Have a brain
- Have a heart
- Have a backbone
- Young live in water
- Adults have lungs

- Walk
- Fertilized internally
- Have vocal sacs
- Fertilized externally
- Are amphibians
- Hatch from eggs
- Young have gills

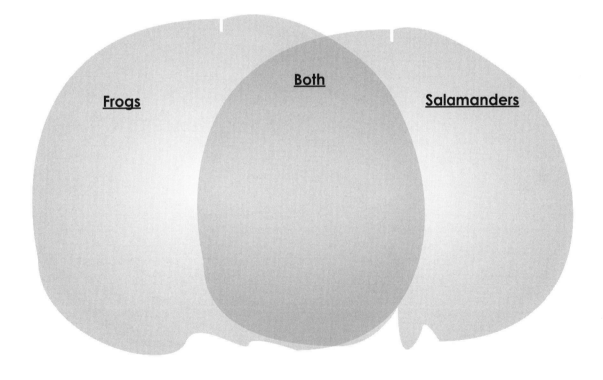

Frogs   **Both**   **Salamanders**

Lily Pad Venn Diagrams

Name(s):_____

Write the descriptive characteristics into the reptiles and amphibians Venn diagram.

- Dry scaly skin or shell
- Replace gills with lungs in adulthood
- Have a heart (some have live young)
- Breathe with lungs at all times
- Have a backbone
- Moist skin
- Most young live in water

- Have a brain
- Young same shape as adult
- Cold-blooded
- Eggs laid on land
- Eggs laid in water
- Young different shape from adults
- Adults have lungs
- No claws

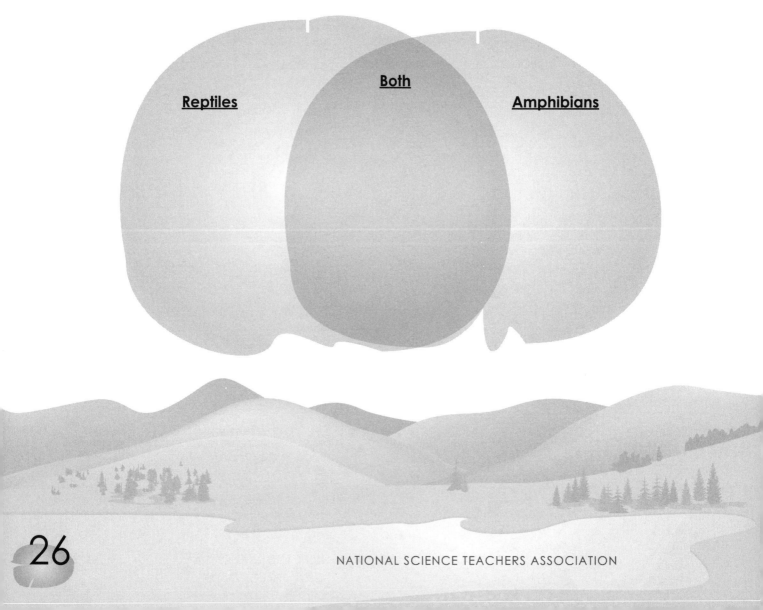

Reptiles   Both   Amphibians

Lesson 5:
Frog Hop Relay Race

Objectives

Students will be able to describe the characteristic movements of amphibians.

Method

Students describe and demonstrate the movement of amphibians and reptiles through a relay race.

Materials

Cones, hula hoops, soft flooring

Background Information

Amphibians have many different ways of moving around on land and in water. Depending on the type of amphibian, it might hop, leap, glide, climb, run, walk, or even burrow. Some frogs have sticky suction-disked fingers that make them excellent acrobats and allow them to climb trees, while others have webbed feet that look like flippers and strong back legs, making them powerful swimmers and leapers. Some frogs can even stretch their webbed feet wide, parachuting or "plopping" into the air and gliding from one area to another (Stebbins and Cohen 1995).

Salamanders may not be able to jump as far or high as frogs, but they can scurry quickly and will leap with surprising grace to avoid capture, as the red-backed salamander does. The caecilians will push headfirst with worm-like

Grade Level: K–2

Subject Area: science, physical education

Skills: analysis, application, description, small-group work

Setting: inside or outside

Lesson Duration: 30 minutes

Group Size: 10 or more

National Science Education Standards, Grades K–4
- **Physical Science:** Position and motion of objects
- **Life Science:** The characteristics of organisms
- **Life Science:** Organisms and environments

contractions into the soil. The head is moved up and down with a ramming action to burrow in their direction of travel (Stebbins and Cohen 1995).

Amphibians will move for a variety of reasons, including capturing prey; avoiding becoming prey; finding a mate or egg-laying location; and finding suitable habitat that has food, water, shelter, and space.

Procedure

1. Begin a discussion with students about how amphibians move at different life stages and how their movements help them avoid predators. Discuss how each movement is an adaptation that allows the animal to survive.
2. Bring students to an open area and ask them to demonstrate the movements made at each life stage.
3. Split the students into two teams so that they can hop like a frog, slither like a salamander, and wiggle like a caecilian in a relay race. The instructor may choose the movement pattern or ask the teams to decide.
4. Instruct teams that only one student from each team will act out the prescribed movement at a time while it's their turn to race through the relay course. When the student returns to the line, the next player may take his or her turn. The first team to have all players participate wins.
5. Players will race through the course around cones or hop into and out of hula hoops and back to the group, where the next student will race, until all students have had a chance to participate.

Reflect and Explain

- Ask students if they could move like any amphibian which would they choose to move like and why?
- Call out different stages of amphibian growth for students to try while the relay race is in motion to test the students understanding of the variety of movement strategies.

Extensions

- Play a game in which students act as frogs that have to cross an open field past a tagger or "predator" without being caught or "consumed." If caught, they instantly become a cattail and must stand swaying in the breeze.
- Create green headbands with a frog picture attached for students to wear in the relay race.
- Allow students to move like aquatic organisms or macroinvertebrates. For example, aquatic worms wiggle, amphipods swim on their sides, and dragonfly nymphs can shoot forward using "jet propulsion" as they push

a burst of water through their gills into their bodies and back into the water body.

- Add additional frog pond animal movements to your relay race. For example, fly like a duck, swim like a fish, and run backward like a crayfish escaping.
- Play a game of charades in which a student imitates a frog pond animal and the other students try to figure out its identity.
- Host a long-hop contest to see which team can hop the farthest.
- Create origami frogs that can actually jump.

Resource Information

Stebbins, R. C., and N. W. Cohen. 1995. *A natural history of amphibians*. Princeton, NJ: Princeton University Press.

Temko, F. 1986. *Paper pandas and jumping frogs*. San Francisco, CA: China Books and Periodicals.

Lesson 6:
Camouflaged Critters

Objectives

Students will be able to demonstrate their knowledge of how camouflage and bright coloration are used for protection and survival of amphibians.

Method

Students will color and hide amphibians to learn about camouflage and use their sense of sight to search them out.

Materials

Amphibian outlines, coloring pencils or crayons, scissors, tape, field guides

Background Information

Many amphibians, particularly those in the tropical rain forests of Central and South America, avoid becoming prey by presenting themselves as brightly colored. Their color acts as a universal warning that their skin secretes powerful poisons that may be toxic enough to kill a predator in seconds. The native people recognized this and used the poisons secreted by the frog's skin on their blowgun darts. However, appearing bright and colorful is not always a guarantee that an amphibian is

Grade Level: K–2

Subject Area: science, art

Skills: description, drawing, identification

Setting: inside or outside

Lesson Duration: 30 minutes

Group Size: no minimum

National Science Education Standards, Grades K–4

- **Science as Inquiry:** Abilities necessary to do scientific inquiry
- **Science as Inquiry:** Understanding about scientific inquiry
- **Life Science:** The characteristics of organisms
- **Life Science:** Organisms and environments
- **History and Nature of Science:** Science as a human endeavor

poisonous, but it does help offer protection from cautious predators (Stebbins and Cohen 1995).

Many of the amphibians in North America appear camouflaged, blending into their native habitats. Upon closer inspection, the coloration of these amphibians may appear different on the top and bottom of their bodies. This sometimes includes a lighter color below the surface of the water that may cause a predator below to not notice their presence. The predator above on land or in the sky may also not notice their presence and miss the opportunity for a meal if the top side of the amphibian blends well into the surrounding vegetation. Many amphibians, particularly salamanders, will also twist their bodies to posture and expose certain colors that help them avoid becoming food or prey or present parts of their bodies, such as the tail, that if removed, can regenerate (grow back) (Stebbins and Cohen 1995).

Procedure

1. Ask students what color they would make their amphibians so they will not be easily found when placed around a pond. If possible, allow students to preview the area, by photo or in person, before deciding on colors. Also ask them what colors they would make their amphibians so they *are* easily found around a pond.

2. Have students color the amphibian outlines in camouflaged or colorful ways and cut them out of the worksheet.

3. Allow half of the group to go hide their amphibians in the available habitat, but be sure to tell them they must be visible without the need to move leaves, rocks, or sticks and to remember where they placed them.

4. Allow the other half of the group to go out to the location where the amphibians were hidden and see how many amphibians they can find in two minutes. You may wish to provide the students finding amphibians with printouts of predators to wear around their neck (raccoon, heron, fish, snake).

5. Repeat the exercise so both groups get to hide and search for amphibians.

Reflect and Explain

- Which amphibians were easier to find? Harder? Why?
- Why did we find more brightly colored frogs?
- Why do some amphibians stand out in their habitats while others blend?

Extensions

- Make a frog or salamander door hanger. Punch two holes at the top of the outline and tie a string through so it can be hung on a doorknob. Color the frog with crayons and colored pencils or make a design with lentils or lima beans and glue. You may wish to write on the door hanger "hop on in."

- Show students two brightly colored frogs, one poisonous and one edible. Wait until the end of class to ask them if they remember which is which. Many students will not remember which one is the poisonous frog. Discuss why the nonpoisonous frog would benefit from looking like the poisonous one (predators might not know the difference, so they won't disturb either).
- Make a shoe box diorama of an amphibian's habitat.
- Play a game of camouflage, offering students different-color costumes, and have them "hide" within sight of the person looking, teaching them to blend in and to look closely for the animals that use camouflage.

Resource Information

Heller, R. 1995. *How to hide a meadow frog and other amphibians*. New York: Penguin.

Stebbins, R. C., and N. W. Cohen. 1995. *A natural history of amphibians*. Princeton, NJ: Princeton University Press.

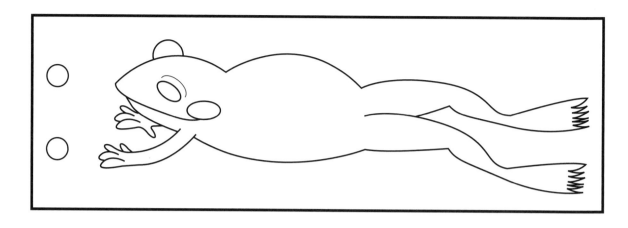

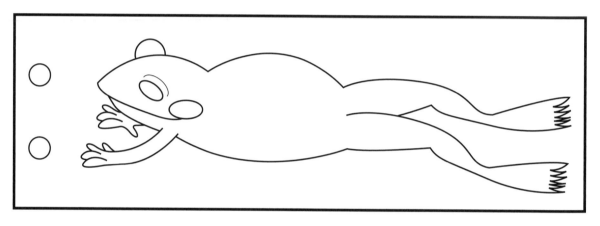

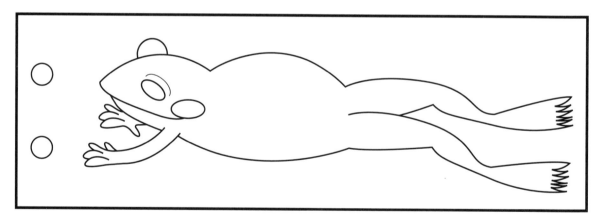

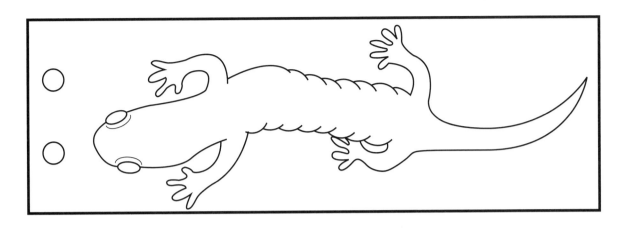

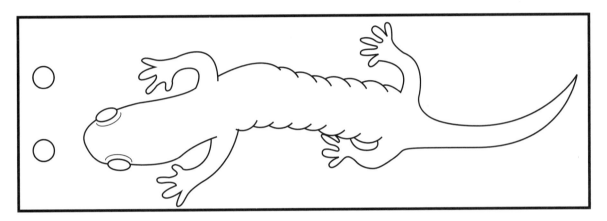

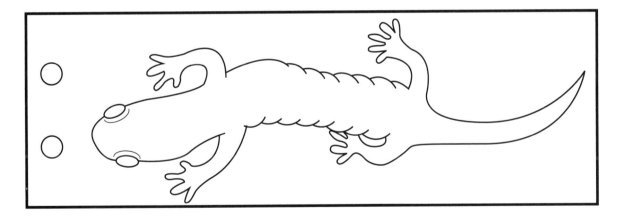

Lesson 7: Amazing Amphibian Migration

Objectives

Students will be able to describe migration as a regular seasonal movement and identify the obstacles amphibians face through natural phenomenon and human behavior as they move through their habitats.

Method

Students will move through an outside area or school hallway to discover picture cards with messages that describe migration challenges.

Materials

Amazing Amphibian Migration Worksheet, laminated set of Migration Cards, clothes pins

Background Information

Amphibian migration refers to the movement of amphibians from upland to the lowland wetland or egg-laying site. Not all amphibians will migrate to find egg-laying locations. However, those that do migrate often require large, uninterrupted habitats to provide connectivity between their wintering upland areas and their breeding areas. Many move when triggered by

Grade Level: 2–4

Subject Area: science, language arts, physical education

Skills: description, identification, inference, small group work

Setting: inside or outside

Lesson Duration: 20 minutes planning, 30 minutes activity

Group Size: no minimum

National Science Education Standards, Grades K–4

- **Life Science:** Organisms and environments
- **Science and Technology:** Abilities of technological design
- **Science in Personal and Social Perspectives:** Characteristics and changes in populations
- **Science in Personal and Social Perspectives:** Changes in environments

warm rains that help keep their skin moist during the migration while also filling the pools at their destination. The reason amphibians may require large, uninterrupted habitats is that during migration amphibians encounter many obstacles. The largest threat to their survival has been habitat destruction and habitat fragmentation. Amphibians have difficulty passing exposed or open stretches of land, such as ball fields or roadways that may cause them to be exposed and vulnerable to predators or automobile traffic.

Believe it or not, amphibians do not migrate all at the same time, but rather they take turns dipping into the pool to lay their eggs to maximize the chances of survival of their offspring. As an example of an early migrant, wood frogs are one of the first amphibians to emerge from hibernation in the early spring to find pools within which they lay their eggs. In February, they will even lay eggs in icy waterways but close to the surface to absorb heat from the sun that speeds up their metamorphosis. Other amphibians such as green frogs and bullfrogs don't lay eggs until summertime. These breeders wait for warmer water that allows for faster metamorphosis and a deeper pool with a greater diversity of prey items available to the young for consumption.

Herpetologists study amphibians and reptiles in several ways. One way has been with the help of citizen scientists working on "bucket brigades" to collect, count, and carry migrating species at specific road-crossing sites. Another is to have volunteers count egg masses in a pool to determine population levels. The information from both studies can then be used to detect the upward and downward trends in amphibian populations. A declining trend may be a sign that the environment is in trouble and additional research or habitat changes are required. Data are often shared by biologists using geographic information systems technology. When collected data are compared from year to year, trends can be observed. This information can be used to protect the habitats of declining populations or to protect habitats of healthy populations that have a stronger chance of survival.

Preplanning

1. Prepare the amphibian migration cards where one side is a photo of an amphibian and the other is an action statement from Figure 7.1 (pp. 42–48).

2. Clip one clothespin to each amphibian migration card. Attach the cards along a trail leading to an amphibian habitat (pond if available). Place the cards slightly out of sight so students only find one at a time.

3. Print copies of the Amazing Amphibian Migration Worksheet for students in grades 3 and 4, and place the worksheets on clipboards so students can write on the trail.

4. Review Safety Practices for Outdoors (p. xxiii).

Classroom Procedure

5. Engage students in a conversation about making plans for a trip. What do you need for a trip? How are you going to get to your desination? What could go wrong on your trip? What could help you get there faster?

On the Trail

6. Explain to students that they are going to be migrating amphibians and that they need to find amphibian migration cards along the trail to find their way.

7. Have students take turns finding the amphibian migration cards. When a card is found, read the instructions on the back. If you have a large group, consider splitting into smaller groups and sending early, middle, and late migrants on the journey. Instruct students not to remove the amphibian migration cards from the trail.

8. After students listen and act out the instructions, continue searching for the next stop. Students may be provided with the amphibian migration worksheet to record the information found on the back of each amphibian card.

9. When you reach your final destination (frog pond), celebrate your successful journey.

Reflect and Explain

- What happened to us as amphibians along the journey?
- Based on the challenges amphibians face, what conclusions can we draw (inferences) about their future? Do they need our assistance to survive, and why?
- How can we help amphibians have a successful journey?

Extensions

- Write a story about the journey of an amphibian.
- Join a local organization to provide assistance to migrating amphibians. Be careful and always watch for traffic and never risk your safety. Offer to create migrating amphibian posters for the organization's road-crossing site.
- Create an educational poster to encourage people to stay off the roads on rainy spring nights and, when amphibians are active, to limit their use of fertilizers and pesticides that can run off and pollute frog habitats.
- Create and use storm drain water stencils with a picture of a frog saying, "Do not dump—this drains to frog habitat"

Resources

Joy Pratt-Serafini, K. J. 2000. *Salamander rain: A lake and pond journal.* Nevada City, CA: Dawn Publications.

Lamstein, S. 2010. *Big night for salamanders.* Honesdale, PA: Boyd Mills Press.

U.S. Department of Transportation: Federal Highway Administration Critter Crossing. *www.fhwa.dot.gov/environment/wildlifecrossings/main.htm.*

Amazing Amphibian Migration Worksheet

Name(s): _____

Amphibian Species: _____

On your journey, what happened to your amphibian?

Summarize the scenario from each card in the space below.

_____

_____

_____

_____

_____

_____

_____

_____

_____

_____

_____

_____

_____

_____

_____

_____

_____

_____

Figure 7.1. Migration Cards

Print and cut cards so the front shows the amphibian.

Winter is ending, and spring is in the air. Begin your migration to the frog pond.

Move ahead.

Frozen ice.

Crawl ahead to find a warmer route.

Watch out for the road!

Crawl ahead on your hands knees.

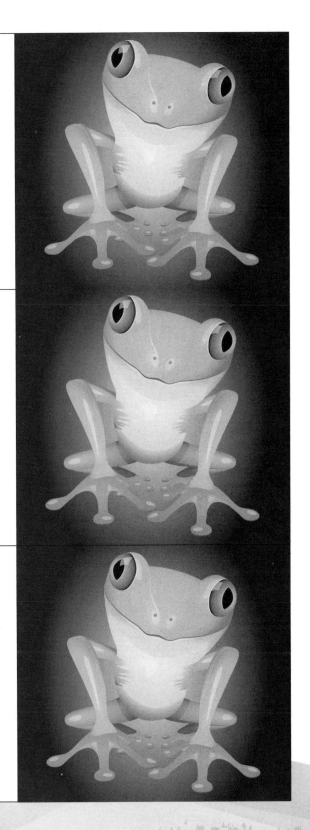

Watch out for the barred owl! It wants to eat you.

Freeze, count to 20, then move ahead.

You escape being hit by a car but lose your tail to a tire.

Slowly roll your shoulders and move ahead.

Your pond has been polluted from fertilizers applied to a local golf course.

Hold your stomach for a count of 10 seconds and groan 5 times.

Herpetologists at the Frog Pond Institute catch you for research. After they apply fluorescent dye to your skin for future identification, you are set free.

Move ahead.

You get picked up by a child who just used bug spray. The chemicals seep into your skin and make you sick.

Rub your stomach 5 times, then move ahead.

It's raining, it's pouring, and you have the opportunity to walk to the frog pond.

Move ahead before the storm ends.

You can't find the spot you came to last year because a new shopping mall has been built on the site.

Walk around in circles looking for a way around.

While traveling at night, you become confused by a street light.

Sit down, count to 20, then move ahead.

You become covered with salt used to melt snow on the roads.

Stick out your tongue and say, "Yuck."

The forest is thick with wet leaves.

Move ahead.

New road construction included tunnel underneath for amphibian crossing.

Move ahead.

Watch out! Raccoon looking for a snack! It wants to eat you.

Zig-zag ahead to avoid becoming prey.

You escape being caught by a turtle but lose a toe.

Pretend to eat an insect to gain some more energy for the journey.

The rain starts turning to snow.

Find shelter before you freeze on the forest floor.

You are able to migrate a good distance because heavy rains keep your skin moist on the journey.

Move ahead to the next space.

Watch out for the garter snake! it wants to eat you.

Freeze and count to 10 before moving ahead.

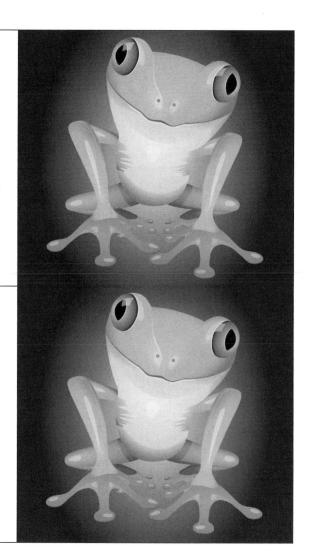

Congratulations!

You have arrived at the frog pond.

These blank cards can be used to write scenarios appropriate to your own habitat.

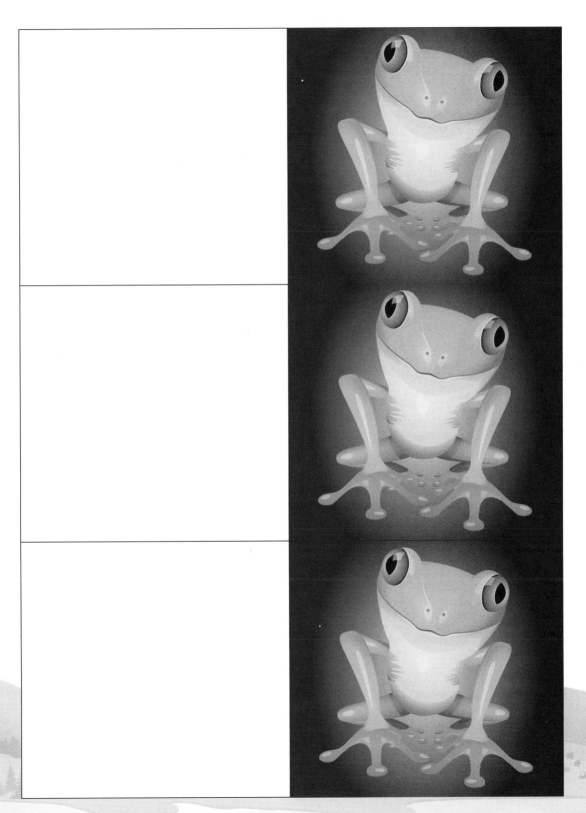

Lesson 8:
Frog Pond Soup

Objectives

Students will be able to describe the components of a frog pond to include living parts (plants and animals) and nonliving parts (water, air, soil, and sunlight).

Method

Students discuss the ingredients that make up a healthy frog pond.

Materials

Index cards, pencil

Background Information

Frog ponds are complex **ecosystems** made up of more than just frogs and pond water. There are many parts that make the habitat livable and eventually attract frogs. There are **biotic** (living) parts, including plants and animals, as well as **abiotic** (nonliving) parts, including water, air, soil, and sunlight. Together these parts allow for a suitable habitat that frogs enjoy.

The living parts of an ecosystem may include frogs, fish, turtles, ducks, clams, crayfish, mosquitoes, dragonflies, beetles, worms, leaches, algae, mosses and liverworts, cattails, lily pads, and much more. Each living thing has a life cycle that includes a beginning (birth for animals, germination for plants), growth, and

Grade Level: 3–4

Subject Area: science, language arts

Skills: analysis, description, small-group work

Setting: inside or outside

Lesson Duration: 30 minutes

Group Size: no minimum

National Science Education Standards, Grades K–4
- **Physical Science:** Properties of objects and materials
- **Life Science:** Organisms and environments
- **Earth and Space Science:** Properties of earth materials
- **Science and Technology:** Abilities to distinguish between natural objects and objects made by humans

death. The nonliving parts of the pond include water, air, soil, and sunlight. Together these parts make up a web of life with interactions—such as animals eating plants and plants getting nutrients from the soil—that allow for life to exist at the pond.

Procedure

1. Tell students that you will make an unusual and unique frog pond soup today. This is not something that will be edible for humans, but rather it is something that could be made from parts of a real frog pond outside. The purpose is to try to determine the types of ingredients that we would need to create a good frog pond or habitat, in this case what we are calling frog pond soup.

2. Take your students on a visit to a frog pond habitat or read a story about pond life. Ask students to list what ingredients they think belong in the frog pond soup. They can begin to do this on the Frog Pond Soup Worksheet.

3. Place pictures of ingredients that may or may not belong in a frog pond into a large coffee can or soup pot. Ask students to pull the pictures out and discuss why they may or may not belong in the pond. Go a step further and discuss what ingredients would be harmful to the pond.

4. As a nonscientific literacy lesson, students may create their own recipes that break down how much of each ingredient should go into the creation of the frog pond soup. Recipes should be written on index cards as if used to actually create a meal or, in this case, a functioning frog pond habitat.

David's Famous Frog Pond Soup

Add 3 beetles, 1 leech, 2 dragonflies, and 1 damselfly,

a pinch of algae, 2 cattails, 3 cardinal flowers,

6 heaping tablespoons of soil, 4 cups of algae,

2 cups of duckweed, 6 cups of water, and 1 large lily pad,

and let sit in the sun until slimy.

Serves 3 to 4 frogs.

Reflect and Explain

- What are some of the components that belong in a frog pond habitat?
- Can frogs live in a pond that only has water, or do they need other living and nonliving things? Why?
- Do you think your ingredients will work 10 years from now? 100? Why or why not?

Extensions

- Create a list of plant and animal species that you would invite to a frog pond pool party.
- Make a pond soup that includes edible representations of the components of a pond.
- With adult supervision, scoop up some water in your hands for a self-conducted sensory introduction to a pond. How does it feel? What does it smell like?
- Create a pond peeper or water scope. Remove both ends of a can and cover one end with clear plastic wrap held on with a rubber band. The covered end can be pressed into the water to see below the surface.

Service Component

- Organize a service learning opportunity where students help with a pond cleanup to remove ingredients that don't belong, like trash, or improve the health of the habitat by adding beneficial native plants.
- Work together to plan and build a frog pond in your backyard or schoolyard.

Resource Information

Reid, G., S. Kaicher, and T. Dolan. 2001. *Pond life: Revised and updated.* New York: Golden Guides, St. Martin's Press.

Frog Pond Soup
Worksheet

Name(s):_____

Create your own Frog Pond Soup.

Ingredient List:

My Frog Pond Soup Recipe

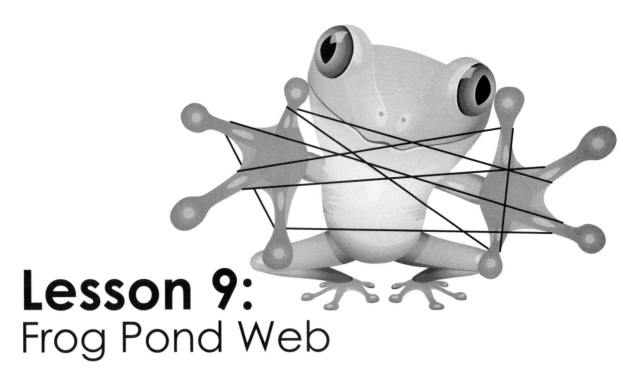

Lesson 9:
Frog Pond Web

Objectives

Students will be able to describe the living and nonliving components of a pond and describe relationships between them.

Method

Students create a list of living and nonliving things found within a frog pond habitat and discuss their relationships to each other through the creation of a simulated web using a ball of yarn.

Materials

Cards of living and nonliving things found at a pond, a hole punch, string, ball of yarn

Background Information

In a frog pond ecosystem, interactions are constantly taking place between both biotic (living) and abiotic (nonliving) things. The study of pond **ecology** is the study of living **organisms** in a pond habitat and their interactions with the nonliving pond environment. Frogs, toads, and sala-

Grade Level: 3–4

Subject Area: science, language arts

Skills: analysis, application, description, identification, inference, small-group work

Setting: inside or outside

Lesson Duration: 30 minutes

Group Size: no minimum

National Science Education Standards, Grades K–4:

- **Science as Inquiry:** Abilities necessary to do scientific inquiry
- **Life Science:** The characteristics of organisms
- **Life Science:** Life cycles of organisms
- **Life Science:** Organisms and environments
- **Science and Technology:** Abilities to distinguish between natural objects and objects made by humans
- **Science in Personal and Social Perspectives:** Characteristics and changes in populations
- **Science in Personal and Social Perspectives:** Changes in environments

manders may all act as **predators** (hunter) and **prey** (hunted) in a frog pond web. They will act as predators consuming energy by eating other plants and animals while trying to avoid becoming prey or energy, playing a key role in both **aquatic** and **terrestrial** ecosystems.

Creating a habitat web allows for students to understand the relationships among things found at a frog pond and how energy moves from the Sun to **producers** (plants) to **consumers** (animals) and becomes recycled by **decomposers**.

Amphibians will play an essential part in the pond ecosystem and often are referred to as **indicator** species. This is because they breathe and drink through **permeable** skin that unfortunately will absorb **pollutants** such as acid rain, pesticides, and other chemicals found in the water. If you notice an absence of amphibians, it may be a warning that the health of the habitat is in jeopardy and should be studied to improve conditions.

Preplanning

1. Print cards provided or create cards of living and nonliving things found at a frog pond. For a list of possible cards, read Procedure step 4.
2. Use a hole punch to make two holes on the top of the cards.
3. Tie a length of string through the holes in the cards so that they can hang around students' necks.

Procedure

4. Begin a discussion about the living and nonliving things found within a frog pond habitat. Students should list their answers on the Frog Pond Discovery Worksheet. Make a list of plants, such as cattails, pickerelweed, duckweed, skunk cabbage, lily pads, and oak trees. Make a list of animals, such as frogs, salamanders, snakes, raccoons, deer, turtles, herons, ducks, fish, dragonflies, mosquitoes, spiders, and worms. Make a list that includes decomposers, aquatic worms, and scavenger beetles. Make a list of nonliving things—the Sun, water, air, and soil and rocks. You may choose to include other nonliving things found at the pond, such as litter. For older students, you may choose to include microscopic life, such as plankton, rotifers, and copepods.
5. Once students understand living and nonliving things, explore the interdependence of living things. Most are dependent on each other in an "eat or be eaten" web of life. Each student should be provided with a picture card or given the opportunity to create a card of a living or nonliving thing found within a frog pond habitat. Allow students an opportunity to brainstorm and research the connections their card has with the cards of other students.
6. Ask students to hang or attach the picture card to themselves so it is visible to the rest of the group when sitting in a circle.

7. Provide lengths of yarn or a ball of yarn to pass around from student to student as they make connections to one another that represent the exchange of energy. The student with the ball of yarn should hold onto the end and pass the ball to another student in the circle, with whom they can create a connection. Examples may include a student with a frog card saying, "I consume mosquitoes," and passing the ball of yarn to the student with the mosquito card. Educators may review and encourage the use of vocabulary words such as *predator, prey, decomposer, carnivore, herbivore, omnivore, consumer, producer,* and *decomposer.*

8. Hand the student with the Sun card a ball of yarn. Have the "Sun" begin by passing energy on to another student in the ecosystem. Continue to have students pass the yarn and be sure to have the students explain how they are connected.

9. When all students in the web are connected, have one student gently tug on the string. When a student feels a tug, they should tug back in response until the web feels tight. Discuss the importance of these connections.

10. Ask students what happens if we remove a connection in the web (others will be affected). What might happen that would remove a connection in the web (habitat destruction, chemicals, harvest)?

Reflect and Explain

- What happens when one connection is lost—can other species survive? Why or why not?
- If an amphibian species declines or goes extinct, what will happen to the prey species they were eating or the predators that ate the amphibian?
- What would happen if all the frogs left the pond?

Extensions

- Complete the Frog Pond Discovery Worksheet before and after visiting a pond.
- Complete the Frog Pond Scavenger Hunt Worksheet while out searching for living things around an amphibian habitat. Have students yell out "Frog Pond Bingo" when they complete a row to win.
- Play a game of "Who Am I?" by inviting a student to the front of the classroom and having them ask yes or no questions to figure out which frog pond animal is posted on their back without getting to look at the picture card.
- Write five descriptive sentences about your frog pond web card without mentioning what living or nonliving thing you have. Read the sentences aloud to see if others can identify what living or nonliving thing you wrote about.

- Create your own habitat web by creating arrows between drawings or pictures of living and nonliving things. Arrows will link predators and prey and producers and consumers to help illustrate the energy flow. An example of a bullfrog habitat web is on page 64.

Resource Information

Ellis, B. 2006. *The web at dragonfly pond*. Nevada City, CA: Dawn.

McKinney, B. 2000. *Pass the energy, please*. Nevada City, CA: Dawn.

Frog Pond Discovery Worksheet

Name(s):_____

Complete indoors.
What do you think you will see at the pond?

Living things: | **Nonliving things:**

Complete Outdoors.
What did you see at the pond?

Living things: | **Nonliving things:**

Do you think any of the living and nonliving things depend on each other? Why?

Frog Pond Scavenger Hunt Worksheet

Name(s): _____

Circle all the living things that you find!

Arrowhead	Cattail	Dragonfly
Duck	Duckweed	Heron
Frog	Snake	Turtle
Pickerelweed	Frog Eggs	Salamander or Newt

NATIONAL SCIENCE TEACHERS ASSOCIATION

Frog Pond Web Cards

Students may use the following cards provided for the frog pond web activity or create their own cards based on something at a pond they have researched.

Example cards that can be worn around the neck.

Sun	Water
Soil/Rocks	Cattail
Pickerelweed	Duckweed
Skunk Cabbage	Oak Tree

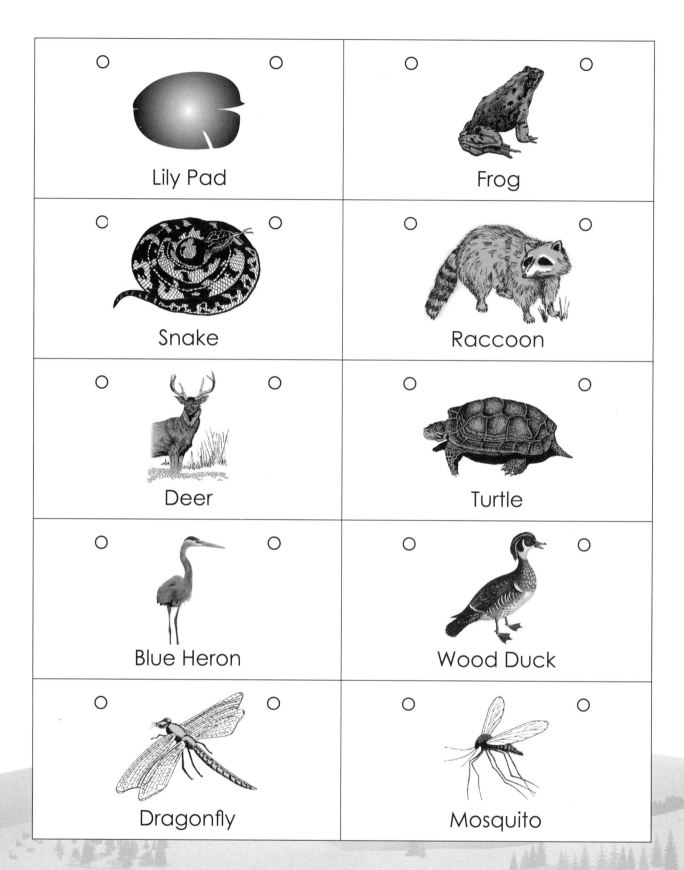

Lily Pad

Frog

Snake

Raccoon

Deer

Turtle

Blue Heron

Wood Duck

Dragonfly

Mosquito

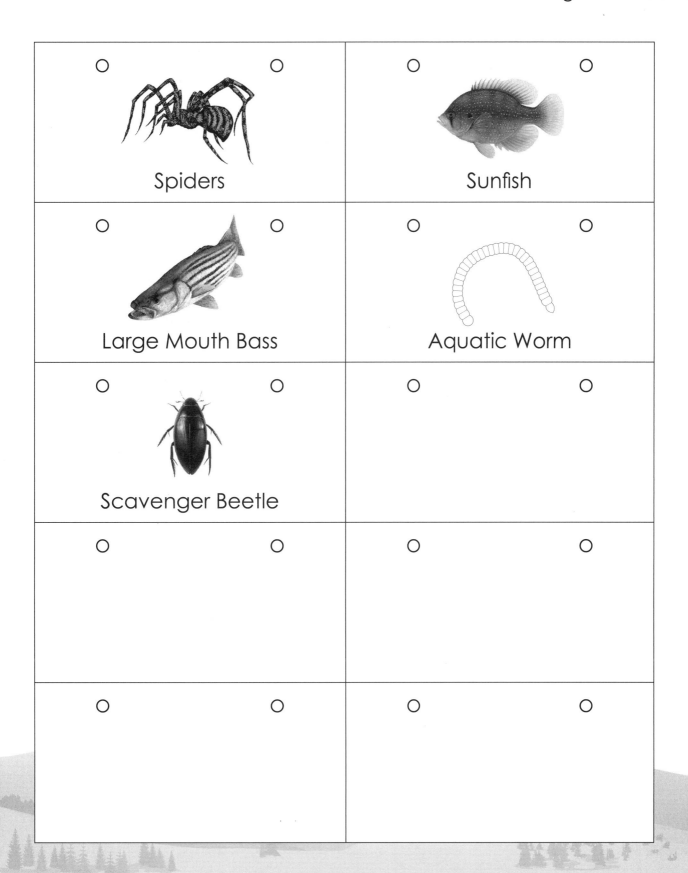

Spiders

Sunfish

Large Mouth Bass

Aquatic Worm

Scavenger Beetle

Bullfrog Food Web

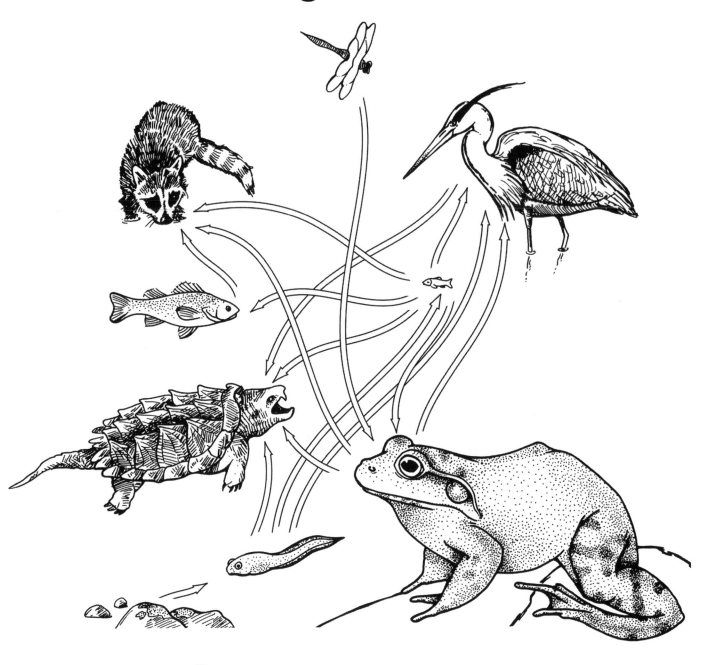

Lesson 10:
Frog Pond Lifeguard

Objectives

Students will be able to determine the health of a frog pond through investigation and inquiry.

Students will be able to describe macroscopic organisms found in pond water.

Method

Students analyze the health of a pond through physical and biological analysis.

Materials

Hand lens, microscopes, slides, covers, pipettes, water samples, clipboards, Frog Pond Lifeguard Worksheet, Aquatic Invertebrate Scavenger Hunt Worksheet, Too Close for Comfort Worksheet, pencils, dip net, viewing containers, white ice cube tray for sorting aquatic invertebrates, forceps, measuring tape, field guides, nonmercury thermometer

Grade Level: 4

Subject Area: science, language arts

Skills: analysis, application, identification, inference, small-group work, synthesis, evaluation

Setting: outside

Lesson Duration: 90 minutes

Group Size: no minimum

National Science Education Standards, Grades K–4:

- **Science as Inquiry:** Abilities necessary to do scientific inquiry
- **Science as Inquiry:** Understanding about scientific inquiry
- **Life Science:** The characteristics of organisms
- **Life Science:** Organisms and environments
- **Earth and Space Science:** Changes in the Earth and sky
- **Science and Technology:** Abilities of technological design
- **Science and Technology:** Understanding about science and technology
- **History and Nature of Science:** Science as a human endeavor

Background Information

Conservation of frog ponds and their surrounding habitat can require local measures. **Citizen-scientists** are often recruited to collect data on natural resources to assist in this conservation effort. With the help of these citizens monitoring our water, we can be quick to notice any impairment—whether natural or influenced by humans—and prevent pollution that may negatively influence the health of our habitat and humans who live within the greater environment.

The physical analysis that students will complete in this lesson is important so that they can describe the characteristics of the pond from visit to visit. This includes a biological analysis that asks students to observe and identify the living organisms within the pond; this may include the micro-organisms and **macro-invertebrates**. These micro-organisms are too small to be seen without a microscope and may include bacteria and protozoa. The pond may also host macro-invertebrates, small animals without a backbone that can be identified by eye or with the help of a magnifying glass and are often found within the leaf litter or substrate at the pond bottom. The types and concentrations of micro-organisms and macroinvertebrates will reflect the water's quality. Insects such as stone fly and caddis fly larvae act as water quality **indicators** because they can't live in polluted water. Finding them indicates the water quality is good and the pond is nonimpacted. If students only find aquatic worms and leeches, the water quality may be poor and negatively impacted. Finding a diversity of living organisms within the pond means a larger, more dynamic food web. If too few macroinvertebrates are found within the pond, students may discuss possible solutions for improving the health of the pond.

Because amphibians can breathe and drink through their skin, they will react early to environmental pollutants because their skin is **permeable** and can absorb toxins. Therefore, their presence alone can provide information about the health of their habitats. If **deformities** begin to appear or amphibians aren't found when and where they once were, this could be a message that something is wrong. As an example, in the mid-1990s, schoolchildren found deformed northern leopard frogs and mink frogs with extra legs (Lannoo 2008). While deformities appear in normal frog populations, the rates in this instance were several times higher than normal. This may have been a result of **pollutants** in the water, parasites, or a multitude of other environmental factors. The students' findings led to continued research efforts to help track down the problem and maintain water quality that supports amphibians.

Procedure

1. Inform students that they are acting as citizen-scientists and will conduct an environmental assessment by collecting data at a frog pond. Explain that they will begin by recording their visual observations before disturbing the pond with nets and other tools.
2. Create groups so that students may work together to accomplish tasks.
3. Review Safety Practices for Outdoors p. xxiii).
4. Provide each group with the Frog Pond Lifeguard Worksheet, the Aquatic Invertebrate Scavenger Hunt Worksheet, and the Too Close For Comfort Worksheet, a clipboard, and a pencil. Inform students that they should work together to complete the worksheets. Tasks should be assigned to individual group members prior to approaching the field site to complete assignments.
4. Approach the frog pond with students and ask groups to spread out and begin their analysis. Stations or areas around the pond can be assigned for specific data collection tasks, with scientific tools available for student use.

Reflect and Explain

- What signs of life did you observe in and around the pond?
- What do your data show? What can you infer from your data about the health of the pond's ecosystem?

Extensions

- Students may attach additional information they collect, such as photographs, sketches, GPS coordinates, and maps.
- Compare results from multiple visits or multiple ponds using graphs.
- Discuss methods for recruiting volunteer citizen-scientists such as students or parents to help assess the quality of frog ponds over time.
- Discuss challenges and methods for protecting amphibians on private property, like incentives for landowners.
- Create a top ten list of reasons we should protect amphibians or how we can protect amphibians.
- Ask students how we can present our findings. To whom? Why?
- Complete a chemical analysis of the frog pond, starting with a pH test.
- Demonstrate the permeability of amphibian skin to pollutants by soaking two hard-boiled eggs, one with the shell and one without, in water with drops of food coloring. Ask the students how they think the two eggs will be affected by the food coloring. Discuss how an amphibian's skin acts like a sponge or like an egg without the shell.

Resource Information

Lannoo, M. 2008. *Malformed frogs: The collapse of aquatic ecosystems.* Berkeley: University of California Press.

North American Reporting Center for Amphibian Malformations. *www.nbii.gov.*

West, L., and W.P. Leonard. 1997. How to photograph reptiles and amphibians. Mechanicsburg, PA: Stackpole Books.

When recording the location of the pond, students may wish to use the following websites for topographic maps or aerial photographs: *www.topozone.com, www.teraserver.com, www.googleearth.com*

Frog Pond Lifeguard Worksheet

Name(s): _____

Date: _____ **Time:** _____

Survey the physical and biological factors of a frog pond to determine its health.

Pool Name/Location: _____

Provide a description of the surrounding habitat and canopy cover.

_____

_____

_____

Did you find any signs of wildlife around the frog pond (living animals, tracks, scat)?

_____

_____

_____

Air Temperature: _____ **Water Temperature:** _____

Turbidity (circle one): Clear   Slightly Turbid   Very Turbid/Murky

Pool Measurements: Length_____ Width_____ Depth_____

Did you find any aquatic organisms in the pond? Use a field guide or the Aquatic Invertebrate Scavenger Hunt Worksheet to identify them.

Type of Organism	# Found

Why do you think the aquatic organisms you found are an important part of the ecosystem?

_____

_____

Aquatic Invertebrate Scavenger Hunt Worksheet

Name(s): _____

Date: _____ **Time:** _____

Search on the pond bottom with nets for the following aquatic invertebrates.

Draw those found but not listed.

Aquatic Worm	Blackfly Larvae	Caddis Fly Nymph
Crane Fly Larvae	Dragonfly Nymph	Leech
Mayfly Nymph	Midge Larvae	Mosquito Larvae
Scuds	Snail	Stonefly Larvae
Water Boatman		

Too Close for Comfort Worksheet

Name(s): _____

Date: _____ **Time:** _____

A Closer Look—My Observations

Place an aquatic organism in a petri dish.

First take a look with a hand lens. What do you see?

Now look with a microscope. What do you see?

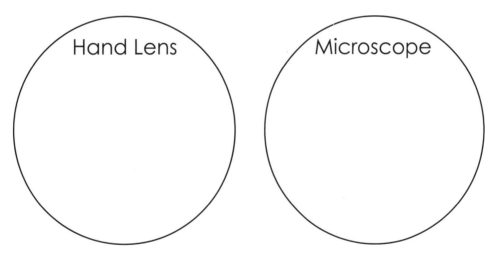

Species Adaptations
Try looking at different aquatic organisms.
Do all the organisms behave in the same way?

In what different ways do they move?

Lesson 11:
Audible Amphibian

Objectives
Students will be able to recognize and identify the calls of frogs and toads. Students will be able to explain that frogs and toads call to communicate.

Method
Students listen, imitate, and identify the calls of frogs and toads.

Materials
Recordings of frog and toad calls, music player

Background Information
Male frog and toad species have a unique **vocalization** or "call" that can be used to identify their species, particularly during the mating season, when they are more audible than they are visible. These male frogs and toads announce their presence with peeps, quacks, chirps, and trills at different times of the day and night to communicate, defend territories, and ideally find a mate. Frog and toad sounds are made as air passes back and forth between lungs and the mouth cavity and through vocal cords, often with the help of expandable balloon-like vocal

Grade Level: 1–4

Subject Area: science

Skills: analysis, application, description, identification, synthesis

Setting: inside or outside

Lesson Duration: 30 minutes

Group Size: no minimum

National Science Education Standards, Grades K–4:
- **Science as Inquiry:** Understanding about scientific inquiry
- **Life Science:** The characteristics of organisms
- **Life Science:** Organisms and environments
- **Science and Technology:** Understanding about science and technology
- **History and Nature of Science:** Science as a human endeavor

sacs that intensify and amplify the volume as they expand and fill, giving the calls greater carrying power (Stebbins and Cohen 1995).

Frogs and toads can not only make sounds but also hear sounds. Sounds and vibrations are captured with their eardrums, called **tympanic membranes**.

Scientists can learn to decipher the ecological soundscape by identifying the sounds of individual species out of a chorus. This can be done through the use of informal tricks such as mnemonic devices and onomatopoeia to help us connect an amphibian and its call. When we listen to frog calls in a chorus, we can determine the diversity of species and monitor to determine the presence of these species from year to year.

Procedure

Part 1

1. Play an audio file of an amphibian chorus or cacophony. Ask students if they have ever listened to the peeps, quacks, chirps, and trills that resonate from an amphibian habitat. Explain that the sounds may be coming from a variety of wildlife, including frogs and toads.

2. Ask students to "repeep" after their instructor to create an amphibian chorus. Play or imitate some of the calls of frogs and toads for your students to hear and repeat. Many calls can be described with a mnemonic phrase. Bullfrogs sound as if they are saying in a deep voice "rr-uum" or "jug o rum," while spring peepers sound as if they are calling a short, loud, high-pitched "peep, peep, peep," and wood frogs sound like ducks quacking "quack, quack, quack."

3. You may choose to make a chorus by splitting students into groups, assigning them an individual frog sound, and waving each group into and out of song.

Part 2

4. Review Safety Practices for Outdoors (p.xxiii).

5. Tell students they are going to let nature do the talking. Students will listen to the sounds of three frogs to see if they can remember their sounds and recognize who is calling when it is played back again.

6. Instruct students to listen to individual calls of a bullfrog, spring peeper and wood frog. Instructors may choose to test their students' abilities to recognize the calls with the Frog Calling Quiz provided or make adaptations for the frog and toad species they wish students to recognize from their local area.

7. Students should write the number of the call they hear (1, 2 or 3) above the amphibian they heard.

Reflect and Explain

- What are the benefits and drawbacks to communicating by sound?
- Why do you think frogs wait until dark to call?
- What was it like to use your sense of hearing to identify amphibians?
- What other animal sounds can you identify by ear?

Frog Calling Quiz

Can you match the amphibian to its call?

Call number:_____   Call number:_____   Call number: _____

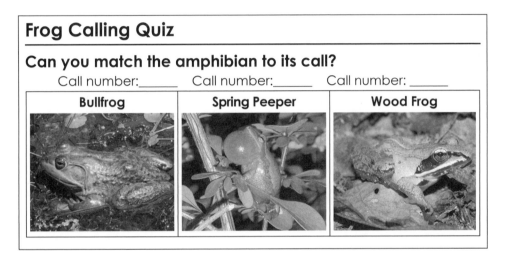

| Bullfrog | Spring Peeper | Wood Frog |

Extensions

- Have students make their own sounds using film canisters and other materials found in the classroom (paper clips, pencils, erasers). See if they can identify a partner while blindfolded, over the noise of the whole group calling to one another.
- Explore an outdoor amphibian habitat and report what you hear to the state division of fish and wildlife or the National Wildlife Federation's "Record the Ribbit" campaign.
- Add additional sounds that you may hear at a frog pond to the test. Examples may include the calls of birds, mammals, and insects.

Resource Information

Davidson, C. 1999. *Frog and Toad Calls of the Rocky Mountains* CD. New York: Cornell Lab of Ornithology.

Eliot, L. 2004. *The Calls of Frogs and Toads* CD. Mechanicsburg, PA: Stackpole.

Ryan, M. 2001. *Anuran communication*. Washington, DC: Smithsonian Press.

University of Michigan Museum of Zoology, Animal Diversity Web, Frog calls. *http://animaldiversity.ummz.umich.edu/site/topics/frogCalls.html.*

U.S. Geological Survey. Frog quizzes. www.pwrc.usgs.gov/frogquiz/index.cfm.

Winer, Y. 2003. *Frogs sing songs*. Watertown, MA: Charlesbridge.

Lesson 12:
Feeding Frenzy

Objectives
Identify the different feeding strategies of frogs

Method
Students simulate the feeding behaviors of amphibians by collecting "prey."

Materials
Colored bottle caps or poker chips, photographs of amphibians using their tongues to catch prey, paper plates (optional)

Background Information
Amphibians have a variety of feeding strategies that allow them to survive and thrive on plants, algae, and bug

Grade Level: K–4

Subject Area: science, physical education, math

Skills: analysis, application, description, identification, inference, small-group work

Setting: inside or outside

Lesson Duration: 30 minutes

Group Size: no minimum

National Science Education Standards, Grades K–4:
- **Life Science:** The characteristics of organisms
- **Life Science:** Organisms and environments

buffets, as well as other prey items that may come across their paths. They may also use their senses of sight, smell, and hearing to detect nearby **prey** (Stebbins and Cohen 1995). Finding food occupies a large percentage of a frog's life and can be difficult as food is distributed throughout **habitats**. Food may only be available seasonally, hidden or not easily detected, and resistant to being caught and consumed. There may also be competition between amphibians and other animals that wish to consume the prey for their own energy and survival.

When frogs hatch from eggs and swim under the water as tadpoles, they are mostly **herbivores** grazing on algae and detritus. When they become adults, they turn into **predators** eating living animals by protracting their long, sticky tongues—which are coated with mucus—to catch and pull in the food item. Some frogs have eyes that will close and pull into the head, helping to push the prey item down the throat and into the stomach. Prey must be swallowed whole because frogs don't do much chewing. Those frogs that have teeth usually only have them on the upper jaw and only use them to crush, hold, and help swallow prey.

Procedure

1. Explain to students that amphibians have different ways of catching their food. Some implement a "sit and wait" technique until the moment prey is within reach, while others actively chase prey to eat. Discuss the benefits and drawbacks to each strategy. If an amphibian is still, it may be camouflaged from predators; but if it is moving, it may give away its location to predators.

2. Split students into two groups. One group will sit still and wait for food items within reach, while the other can move to find food items. The sitting group can be given paper plates colored green to resemble lily pads that they must place on the ground and remain in contact with at all times.

3. The instructor should then toss bottle caps or poker chips into the playing area and allow students to collect as many as they can.

4. Determine which group of amphibians found more food, graph the data and discuss.

Reflect and Explain

What are the advantages and disadvantages of each feeding strategy?

Extensions

- Play "Frog or Fly" in a large open space. Pick one student to start as a frog in the middle of the field, gym, or open space while the other students start as flies on the sidelines. The frog calls out "ribbit" and all the flies must cross the playing field without being tagged or "eaten." When a fly is caught, it will become a frog, but if the frog misses a fly during a round, it will turn back into a fly.

- Use a nighttime lightbulb (often red) for sale at most pet stores or a flashlight covered with red cellophane to watch nocturnal amphibians feed.

- Make an amphibian mask and provide green blowouts (found at party stores or novelty stores) that can be used with Velcro to catch fake flies.

Resource Information

Deban, S. 2009. Amphibian feeding movies. University of South Florida. *http://autodax.net/feedingmovieindex.html*

Faulkner, K. 1996. *Wide mouthed frog*. New York: Dial.

Lescroart, J. 2008. *Icky sticky frog*. Atlanta, GA: Piggy Toes Press.

Stebbins, R. C., and N. W. Cohen. 1995. *A natural history of amphibians*. Princeton, NJ: Princeton University Press.

Lesson 13:
Salamander Smell

Objectives
Students will be able to describe how and why amphibians use their sense of smell for odor communication.

Method
Students use film canisters to follow scent trails that end at a pond or alternative final destination.

Materials
Film canisters, cotton balls, scented oils

Background Information
Evidence supports the use of scent smell or olfaction in the orientation and movement of amphibians toward

Grade Level: 2–4

Subject Area: science

Skills: analysis, application, description, identification, inference, small-group work

Setting: inside or outside

Lesson Duration: 30 minutes

Group Size: no minimum

National Science Education Standards, Grades K–4:
- **Life Science:** The characteristics of organisms
- **Life Science:** Organisms and environments

breeding sites or home areas and in their ability to determine suitable habitat and capture prey (Duellman and Trueb 1986). This is known because researchers have experimented with blindfolding salamanders and have observed them find their way around. Despite the salamanders' lack of vision, they were able to use their sense of smell to navigate. Researchers have also observed blind cave salamanders locate prey so precisely by smell that they were able to shoot their tongues out in total darkness to catch a meal (Stebbins and Cohen 1995).

Some amphibians are even said to have internal "homing devices" that allow them to find the same breeding pool in which they were born each year. It is thought that they recognize the odors in and around the pool so precisely that they can return to it from an upland habitat when it is time to mate. The amphibian's ability to use its sense of smell among other sensory information allows it to navigate and find precisely its desired destination.

Preplanning

1. Place cotton balls in film canisters and add a small amount of essential oil to be absorbed into the cotton ball. Choose scented oils that students may be familiar with and that are distinct enough to tell apart, such as peppermint and lemon. Poke a hole in the cap of each canister.
2. Prepare another set of film canisters with a different-smelling essential oil.
3. Lay the canisters along a trail so students must follow one scent to one destination and the other scent to another destination.

Procedure

4. Ask students how they think salamanders find their way around. What senses do they have that we share? (They can see, hear, touch, smell, and taste.) Which of these senses helps them find their way around? Explain to the students that salamanders can see with their eyes but can also use their sense of smell to navigate.
5. Divide students into small groups and tell them they will have to use their sense of smell to find a final destination, like a pond, as an amphibian would.
6. Allow them to begin at a preset location and instruct them to follow a trail lined with a specific scent provided to them as the first scented film canister.
7. Place mixed scents along a trail that require students to check and compare so they don't get off the trail. They should only follow the scent they were given at the beginning of the trail. Students should not open and touch the scented cotton ball; they should just smell at the opening.

Reflect and Explain

- Was it easy or hard to follow the scent trail and find the final destination? Why?
- Would you still be able to survive without certain senses you've grown to rely on? Which ones? How?
- What senses do animals use to travel with other than scent or odor marking?

Extensions

Replace scents in the canister with beads, paper clips, marbles, pennies, rice, and other materials that allow for this activity to be performed with hearing rather than smelling. Remember to set up the canisters in pairs with equal amounts.

Resource Information

Duellman, W. E., and L. Trueb. 1994. *Biology of amphibians.* Baltimore, MD: The Johns Hopkins University Press.

Ryan, M. 2001. *Anuran communication.* Washington, DC: Smithsonian Press.

Stebbins, R. C., and N. W. Cohen. 1995. *A natural history of amphibians.* Princeton, NJ: Princeton University Press.

Lesson 14:
Frog Pond Poetry

Objectives
Students will be able to write poetry about amphibians.

Method
Students create poetry using descriptive words from their observations.

Materials
Clipboard, paper, pencil, Frog Pond Poetry Worksheet

Background Information
Writers from around the world and various cultures have shared their

Grade Level: 3–4

Subject Area: science, language arts

Skills: analysis, description

Setting: inside or outside

Lesson Duration: 60 minutes

Group Size: no minimum

National Science Education Standards, Grades K–4:
- **Life Science:** The characteristics of organisms
- **Life Science:** Organisms and environments

impressions of amphibian habitats, from Matsuo Basho's "Spring Days" in 1686 to Robert Frost's "Spring Pool" in 1928. As one walks around an amphibian habitat, there is much to observe and describe using poetry. The environment has inspired many writers and can be used to build on the vocabulary of each student. Provided are two forms of poetry that give the students structure for their ideas and allow for the use of descriptive words that help the reader understand the location and feeling felt by the writer without actually being there.

Procedure
1. Share aloud with the students poems from the reference materials.
2. Discuss the structure of a Japanese haiku, a type of poem traditionally written about nature, with the students. Explain that the first line has five syllables, the second line has seven syllables, and the third line has five syllables.

3. Discuss the structure of a cinquain with students. Explain that line one should include only one word to name the subject (noun). Line two should include two words to describe it (adjectives). Line three should include three words of action about the subject. Line four should include a four-word phrase about the subject, and finally, line five should include one new word that names the subject (noun).
4. Review Safety Practices for Outdoors (p. xxiii).
5. Bring students outside to observe a frog pond.
6. Instruct students to brainstorm words before writing. They can fold a piece of lined paper in thirds and title each third as nouns, adjectives, and adverbs. The group members may compare and share words to extend their vocabulary.
7. Write a haiku, cinquain, or both near an amphibian habitat using the Frog Pond Poetry Worksheet, pencil, and clipboard.
8. Read aloud the finished poems or have an adult read them aloud expressively.

Reflect and Explain

- Does your poem conform to the standards of the haiku or cinquain?
- Did the words you chose describe the physical habitat?

Extensions

- Instruct students on how to create acrostic poems, couplets, quatrains, longer rhymes, free-form poems, shape poems, and even limericks.
- Submit poetry to a local newspaper or publication so student work and learning experiences are shared with a larger community.
- Pair poems with amphibian artwork for display.

Resource Information

Florian, D. 2001. *Lizards, frogs, and polliwogs.* New York: Harcourt Brace.

Sidman, J. 2005. *Song of the water boatman and other pond poems.* New York: Houghton Mifflin.

Frog Pond Poetry Worksheet

Name: _____

Think of what you would like to remember most about visiting the pond.

Write a nature poem about your pond visit using the haiku format explained below. Haiku is a traditional Japanese form of poetry about nature. Unlike some forms of poetry, haiku does not rhyme.

	Example
Line 1: 5 syllables (or "beats")	Frogs are all around
Line 2: 7 syllables	Camouflaged in the big pond
Line 3: 5 syllables	Happy in the rain

_____

_____

_____

Write a poem about your pond visit using the following format, called a cinquain.

	Example
Line 1: One word to name the subject (noun)	Pond
Line 2: Two words to describe it (adjectives)	Small, green
Line 3: Three words of action about it	Wet, slimy, bubbling
Line 4: Four-word phrase about it	Where the frogs are
Line 5: One new word that names the subject (noun)	Pool

_____

_____

_____

_____

_____

Lesson 15:
Ribbiting Discoveries in the Lily Pad Paper

Objectives
Students will write a journalism-style story describing the events at a frog pond.

Method
Students research interesting things about amphibians or things found around a frog pond to create a news article for the Lily Pad Paper.

Materials
Lily Pad Paper worksheet, pen or pencil

Background Information
Newspapers have reported on many amphibian discoveries and written headlines to cover amphibian news. Many amphibian-related articles are freely available online and should be used to introduce the students to

Grade Level: 3–4

Subject Area: science, language arts, media, art

Skills: analysis, application, description, drawing, identification, inference

Setting: inside and outside

Lesson Duration: 60 minutes

Group Size: no minimum

National Science Education Standards, Grades K–4:
- **Life Science:** The characteristics of organisms
- **Life Science:** Organisms and environments
- **Earth and Space Science:** Changes in the Earth and sky

this lesson. The cases below may be researched as examples of amphibian news. Discussion may follow to check for comprehension:
- In 1995, students from Henderson, Minnesota, discovered that a large number of northern leopard frogs captured at a local farm pond had deformities of some kind.
- In 2006, scientists discovered a species of tree frog in the mountains of Panama. The frog, known as Rabb's fringe-limbed tree frog, is worthy of print because it has large webbed feet that look like mitts and allow the

frog to leap and glide to the ground from a height of 30 feet.
- In 2009, more than 200 million frogs were caught for human food consumption, leading researchers at the University of Adelaide to report frogs as food to be a possible cause of extinction for some species.

Procedure

1. Tell your students they are starting a newspaper and will take on roles and responsibilities. You need writers to create feature stories, report weather, make comics, obituaries, and so on.
2. The theme of the Lily Pad Paper will be amphibians or life around the frog pond. Provide basic research materials for students to use in writing their stories.
3. The paper should offer a variety of elements and may be fun but should be checked for accuracy. Examples of headlines might include: "Big Green Eating Machine Swallows Dragonfly," "Slithering Serpent Swallows Friendly Frog," "Lily Pad Capsizes From Overfed Frog," "Tadpole Graduates to Frog," "Frog Slips Away From Second Grader." Or articles may include interviews with classmates about their first frog experience, information about frog pond habitat cleanups, interviews with frog pond species, or a story in chronological order about the pond through a day, week, or season.
4. You may choose to allow students to work in pairs or groups to combine ideas and talents.
5. Students may use the Lily Pad Paper worksheet for the assignment. When assignments are complete, put them into a news format that can be copied for distribution.
6. Instruct students to provide editing and feedback on one another's writing.

Reflect and Explain
- Discuss accuracy within the stories; Could the events happen?

Extensions
- Post student work on a class or nature center website.
- Invite a local reporter or environmental communicator to speak about what it takes to get articles published.
- Write letters to the editor of a local newspaper about amphibian conservation.
- Write a 140 character explanation known as a "tweet" about amphibian news.
- Create a bumper sticker, bookmark, poster, or T-shirt as a fund-raiser for amphibians.

Resource Information
Pratt-Serafini, K. J. 2000. *Salamander rain: A lake and pond journal.* Nevada City, CA: Dawn Publications.

Date: _____

Lily Pad Paper

Headline: _____

Writer: _____

Lesson 16:
Seasonal Discoveries Journal

Objectives
Students observe broadly and focus closely to discover seasonal changes.

Method
Students act as naturalists spending time documenting their observations by writing and sketching in journals over repeated sessions that span changing seasons.

Materials
Clipboards, paper, colored pencils, markers, rulers, magnifiers

Background Information
Every day of every season brings new discoveries to write about. Naturalists have long used journals to record observations about the timing of life cycle events such as the times when seasons change, flowers bloom, birds migrate, and leaves change color in autumn. The study of this timing is known as **phenology**. Around a frog pond there will be new observations to describe on a daily, weekly, monthly, and yearly basis. These observations can inform us of changing trends over time.

In winter, there may be little to observe at first glance because many frogs dig themselves into the mud at the pond bottom. Others crawl into animal tunnels or rotting logs. Amphibians are cold-blooded, or **ectothermic**, and rely primarily on external sources of heat to regulate their body temperatures, as they do not stay

Grade Level: 3–4

Subject Area: science, language arts

Skills: analysis, description, drawing, identification, inference

Setting: outside

Lesson Duration: 30 minutes

Group Size: no minimum

National Science Education Standards, Grades K–4:
- **Life Science:** The characteristics of organisms
- **Life Science:** Organisms and environments
- **Earth and Space Science:** Changes in the earth and sky

constant like ours. As a result, when they encounter cold temperatures, as when winter approaches, they must enter a state similar to hibernation known as **torpor**, where their body is in a short-term state of decreased activity. In fact, some frogs, such as wood frogs, become "frogsicles," but they make lots of sugar glucose, which they pump into their blood flow, preventing the cells from being pierced by tiny crystals of ice. They will remain motionless without food as their body comes to an almost complete standstill. Stored fat within their bodies will help provide what energy they need until the next season. Come spring, when melting snow, ice, and rain fill the ponds and the ground warms and the sun shines, their bodies will return to an active state. The frogs will then undergo their most energy-intensive activities, calling and mating. Eggs will be laid and quickly hatch into tadpoles that continue to metamorph into their adult forms.

Allow your students the time to make exciting observations and detail the natural environment around them so they can record in their own journals. Student observations may include sightings of turtles, snakes, fish, and insects. If given time, students may be able to observe the entire life cycle of an amphibian from egg to adult. They can also observe the landscape and document details about a frozen pond in winter that melts in spring, becomes surrounded by budding trees, and eventually is covered in vegetation in the summer.

Procedure

1. Inform students that they will maintain a journal for an extended period of time to record observations in a natural setting. Try to begin journal writing at the beginning of a school year or camp to maximize the opportunity to observe change.
2. Students should be instructed to use the provided Seasonal Discoveries Journal Worksheet or to record similar information in their own journals. It is important that past journal pages are available for students to revisit their prior entries so that they may remember past visits and notice the subtle changes that may have taken place since that visit.
3. Explain to students that the entries should reflect open exploration of the broad environment as well as closely focused observations, such as that of a frog in its different stages of metamorphosis.
4. Harness the creativity of students through journaling of observations and sketch findings. Have them accurately record their findings so that experiences can be shared with a broader audience.

Reflect and Explain

- Ask students what they have learned in the process of observing and recording.
- Ask students what changes they predict would happen at different time scales if more sessions were made available for journal writing.

Extensions

- Read students the story "I'm in Charge of Celebrations" to help them understand that one can expect to see and celebrate seasonal changes.
- Count the total number of frogs at the pond on the same day and time each week to compare from week to week or year to year. Graphs can be made from the collected data.
- Record observations with dates on a chart.

Resource Information

Baylor, B. 1995. *I'm in charge of celebrations*. New York: Aladdin.

Carroll, D. 1999. *Swamp walker's journal: A wetlands year*. Boston: Houghton Mifflin.

Morrison, G. 2002. *Pond*. Boston: Houghton Mifflin.

Pratt-Serafini, K. J. 2000. *Salamander rain: A lake and pond journal*. Nevada City, CA: Dawn Publications.

Stewart, M. 2009. *Under the snow*. Atlanta, GA: Peachtree.

USA National Phenology Network. *www.usanpn.org*.

Seasonal Discoveries Journal Worksheet

Name: _____

Date: _____ **Time of Day:** _____

Place: _____ **Weather:** _____

Record your observations below.

NATIONAL SCIENCE TEACHERS ASSOCIATION

Lesson 17:
Herp, Herp, Hooray

Objectives
Students will develop and explain their own conservation ideas.

Method
Students will work in teams to create their own amphibian conservation plans.

Materials
Paper, pencils, pens, colored pencils, computer with access to the internet, Amphibian Conservation Pledge of Action worksheet

Background Information
Scientists are working to protect amphibians through a variety of **conservation** programs. Conservation efforts are essential because amphibians have been in decline for a variety of reasons, such as from habitat loss, pesticide use, fire, disease, invasive species, climate change, road mortality, and even harvest.

Grade Level: 4

Subject Area: science, language arts

Skills: description, analysis, evaluation, synthesis, public speaking

Setting: inside or outside

Lesson Duration: 60 minutes

Group Size: no minimum size

National Science Education Standards, Grades K–4:
- **Science as Inquiry:** Understanding about scientific inquiry
- **Science and Technology:** Abilities of technological design
- **Science and Technology:** Understanding about science and technology
- **Science in Personal and Social Perspectives:** Science and technology in local challenges
- **History and Nature of Science:** Science as a human endeavor

Scientists have made many contributions to amphibian conservation in the past, and projects exist to protect amphibians all around the world. In 2009 in Cordillera Central, Colombia, where the highest concentration of coffee is produced in South America, the Ranita Dorado Amphibian Reserve was created to protect and provide habitat for a recently discovered small golden poison frog (*Ranitomeya tolimense*) and Swainson's poison frog (*Ranitomeya doriswainsonae*). This reserve is the first of its kind and will attempt to save the last remaining mountain forests where the endangered frogs still live before the surrounding land is developed for agriculture. The people who live around the preserve will gain a new frame of reference, viewing the habitat as an ecological resource that brings economic growth through ecotourism. Local people may have opportunities to share their sensitive habitats, culture, and artistic representations of amphibian art as travel guides and merchants.

In Puerto Rico, crested toads have been collected for zoos and aquariums to breed in **captivity,** and tadpoles are released into the wild where ponds have been built to replace destroyed habitats.

In the northeastern states of North America, there are multiple projects that protect amphibians from road mortality. Projects include road closings to allow amphibians to cross in rainy weather when they are most active and the installation of culverts with a technological design that allows amphibians to pass underneath the road safely. In addition, some of these states have guidelines for certifying seasonal ponds known as **vernal pools** that act as egg-laying locations for many sensitive species of amphibians.

Shared concern about the environment and amphibians has led to varied projects around the world. Successful projects most often contain a framework that allows for a sequence of problem-solving steps. Scientists will state the problem, in this case the decline of amphibian species, and follow up by designing an approach and solution. If reviewers judge that the plan created appears to have the ability to solve the problem, it may be put into action. Once in action, the project will be followed up with an evaluation to determine its success. Students may chose to follow through with a service-learning project. The projects may include civic involvement that can help develop a sense of self-efficacy and improve academic achievement, social skills, and civic-mindedness.

Procedure

1.   Inform students that you need their help to protect amphibians. Explain that they will design a proposal for an amphibian conservation project. Students may use their creativity to design and describe projects on their own or be provided with ideas to develop, such as habitat protection, breeding programs, and technological designs such as new road tunnels. They may even choose to create an educational campaign that focuses people's attitudes and actions toward amphibian conservation.

2.   Break students into brainstorming groups and ask them to explore and investigate ways they could prevent the problem of amphibian decline

and protect amphibian species. You may choose to make groups responsible for creating plans for specific amphibians within your region that are threatened or endangered.

3. Student groups should reflect and explain during a preliminary presentation to other students. The students listening to the ideas of their peers should be given the opportunity to provide constructive feedback and questions. This follow-up will encourage and offer opportunities for further investigation that can lead to the creation of improved designs.

4. When groups have improved and completed their designs, they will present their plans. You may choose to invite guests who act as a panel with the responsibility of deciding on the best plan to put into action. The panel should be sure to thank all students for their plans that may help create a lifeline for sensitive amphibians to survive.

5. Ask students to complete the amphibian conservation pledge of action provided.

Reflect and Explain

- Do you think your plan will work to protect amphibians in five years? One hundred years?
- Are any other parts of the habitat or community negatively affected by your plan? How so?
- What would happen if more scientists or citizens worked to protect amphibians?
- Follow through with a plan to help amphibians in your own community.

Extensions

- Invite a biologist or professor working toward amphibian conservation to speak to your group.
- Have the group describe and list what they would expect an amphibian biologist to carry into the field.
- Ask students to draw or write an essay that illustrates how they might personally care for amphibians.
- Ask students if they believe it is our responsibility to help amphibians. Discuss why or why not?

Resource Information

Franke, V. (Producer). 2008. *Rattlers, Peepers, and Snappers* motion picture. United States: Peregrine.

Amphibian Conservation Pledge of Action

I, _____,

pledge to help amphibians and their habitats.

The issues I am concerned about are

_____

_____

_____

_____

The steps I will take to "Hop Into Action" are

_____

_____

_____

_____

Lesson 18:
Frog Pond Choices

Objectives
Students will examine and discuss their own values around real-life environmental scenarios that involve amphibians.

Method
Students discuss and evaluate complicated decisions that take place around a frog pond habitat.

Materials
Frog Pond Choice Scenarios cards

Background Information
Due to the sensitive nature of amphibians, there are many scenarios that can take place around an amphibian habitat that can be harmful to the local populations. These scenarios require problem identification followed by discussion and explanation of tasks that may alleviate or solve the problem. As citizens and members of a community, we often have to make decisions that may even create constraints for ourselves or others, such as time and financial costs. Before these decisions are made, they often need to be discussed

Grade Level: 4

Subject Area: science, language arts

Skills: analysis, small-group work, public speaking, synthesis, evaluation

Setting: inside or outside

Lesson Duration: 30 minutes

Group Size: no minimum size

National Science Education Standards, Grades K–4:
- **Life Science:** Organisms and environments
- **Science and Technology:** Abilities of technological design
- **Science and Technology:** Understanding about science and technology
- **Science in Personal and Social Perspectives:** Science and technology in local challenges
- **History and Nature of Science:** Science as a human endeavor

within a group and followed up with presented proposals that are evaluated by members of the community. In this activity, students must work in teams to propose solutions to fictitious dilemmas that occur at a frog pond.

Procedure

1. Provide each group of students with a card from Figure 18.1 that offers a scenario and a variety of options that may or may not offer a solution.
2. Allow individual groups to discuss options and come to a consensus on a solution.
3. Invite each group or a representative from each group to read a card and discuss the agreed-upon choice.

Reflect and Explain

- Are decisions always easy and clear to make?
- What can you do for assistance in finding solutions?
- What are the pros and cons to the decision(s) you made?

Extensions

- Have students create their own frog pond choice scenarios to present to the group.
- Ask students to create their own educational flyer illustrating how we can reduce our impact on amphibian habitats to educate those who may be unaware of amphibians' needs.
- Organize a "Frogfest" or "Sala-meander Walk" to promote awareness about amphibians in your own neighborhood.

Resource Information

A Thousand Friends of Frogs "Frog"-quently Asked Questions. *www.hamline.edu/ cgee/frogs/science/faq1.html.*

Figure 18.1. Frog Pond Choice Scenarios

Scenario #1

There is a new road to the new school in town, and frogs are getting run over as they attempt to cross it. What should we do?

1. Close the road when the frogs are crossing and cancel school for the day.
2. Build an underground tunnel for the amphibians to cross.
3. Do nothing and allow them to get crushed.
4. Other:

Scenario #2

Someone threw trash around the pond over the weekend, and now a cleanup will need to be planned. What should we do to prevent future disturbances?

1. Build a fence around the pond and create a locked entrance.
2. Create a sign that explains the importance of the pond so visitors recognize its value.
3. Place a sign that reads "Take Only Pictures, Leave Only Footprints."
4. Install a web cam that monitors the pond and uploads photographs every 10 seconds to the class website.
5. Other:

Scenario #3

The school sprays chemical herbicides on the ball fields within 100 feet of the frog pond. The chemicals could potentially harm the amphibians within the pond. What should we do?

1. Ask the principal to limit pesticides on school grounds.
2. Find pesticides that do not harm amphibians.
3. Other:

Scenario #4

Some students want to take frogs found in the schoolyard habitat home as pets. What should we do?

1. Allow students to take frogs.
2. Allow one student to take a frog and bring it back for another student to adopt over the next weekend.
3. Do not allow students to take frogs from the pond.
4. Other:

Scenario #5

Your friend has a frog he got at the pet store and he does not want to care for it anymore. What should he do?

1. Release the frog into a nearby pond.
2. Ask in school if another student would like to adopt the frog.
3. Return the frog to the pet store even without a refund.
4. Keep the frog in the class as an animal ambassador to be used for student learning.
5. Other:

Scenario #6

Your dog bites into a toad and gets sick from the poison in its parotid glands. What should you do?

1. Kill the toad and others like it.
2. Keep the dog inside.
3. Remove toads and put them in your neighbor's yard or a local park.
4. Do nothing.
5. Other:

Lesson 19:
Frogville Town Meeting

Objectives

Students identify and describe the points of view of different stakeholders in a fictitious community dilemma. Students develop an understanding of how communities are able to solve problems despite citizens with different points of view.

Method

Students take on the roles of various stakeholders in a town meeting.

Materials

Situation statement and information cards; skit materials such as name tags, suit ties (clip-on), construction worker hats, and other costumes

Background Information

As citizens in a **democracy** we have the right to identify, investigate, analyze, and share our values and beliefs

Grade Level: 4

Subject Area: science, language arts, civics

Skills: analysis, application, inference, small-group work, public speaking

Setting: inside

Lesson Duration: 120 minutes

Group Size: 20

National Science Education Standards, Grades K–4:
- **Science as Inquiry:** Abilities necessary to do scientific inquiry
- **Science as Inquiry:** Understanding about scientific inquiry
- **Life Science:** The characteristics of organisms
- **Life Science:** Life cycles of organisms
- **Life Science:** Organisms and environments
- **Science and Technology:** Abilities of technological design
- **Science and Technology:** Understanding about science and technology
- **Science in Personal and Social Perspectives:** Characteristics and changes in populations
- **Science in Personal and Social Perspectives:** Changes in environments
- **Science in Personal and Social Perspectives:** Science and technology in local challenges
- **History of Science:** Science as a human endeavor

on issues that may cause conflict in our communities. The consequences of issues can be minimized when the **stakeholders** or people with varying perspectives take time to evaluate the potential impacts of the issue, recognize tensions that occur between the different beliefs, and work together or within the civic system to find an agreement or a solution. In this lesson, students are split into stakeholder groups involved in a sensitive land-use development conflict where senior homes may be built within the wetland habitat of an endangered northern cricket frog. Ideally, the stakeholders will be able to identify proposed solutions to the issue and discuss arguments for and against them to come to a consensus where an alternative approach is agreed upon. This lesson is meant to help learners develop confidence in their effectiveness as citizens and recognize that multiple points of view are acceptable.

Procedure

1. Begin a discussion about decision making by asking students how they solve problems when people don't agree. For example, find a mediator to decide the solution, play "rock, paper, scissors," or negotiate and choose a solution together using collaboration.
2. Provide copies of the situation statement to be read aloud in class.

Frogville Situation Statement

The town of Frogville is a small, friendly, forested town of approximately 5,000 people. It has been approached by the Senior Housing Corporation about the construction of a planned adult community of 210 apartments in town. If built, the senior housing would provide homes for approximately 400 seniors. The senior homes would be built on 95 wooded wetland acres next to the lake in town, within five miles of the local school and on a new road. The proposed site is an undeveloped wooded wetland where some people say, "Mosquitoes will eat you alive" and others say, "it is full of interesting plants and animals." The Frogville Conservation Coalition is opposed to the construction of the facility because it is worried about the traffic and the noise, water and air pollution, and particularly the northern cricket frog (*Acris crepitans*). The frog is listed by the New York State Department of Environmental Conservation as an endangered species.

Senior Housing Corporation

The Senior Housing Corporation has proposed the development of a senior housing complex that will create housing for 400 seniors. They maintain the construction will meet the rules and regulations of the town and have little impact on the environment. The corporation also says the construction will bring jobs to the community.

Senior Citizens Group

The senior citizen group wants new housing that meets their needs and provides for their interests. They would like to have their own housing so they do not feel like a burden living with their children. They do not care where the development is built so long as it is built near the community.

Frogville Planning Board

This group is responsible for ensuring that guidelines of building laws are followed and minimal impact is made on the environment. They are unsure of the potential impact of the housing development and have many questions to ask before they inform the town council that they approve the building plans.

Frogville Chamber of Commerce

This group repesents local businesses in Frogville and has members that support both the construction of the senior housing and the protection of the habitat. There are many businesses that would benefit from the construction and potential increase in traffic, while others benefit from the tourism of people attracted to the lake for its beauty. Students in this group will represent the different business owners on both sides of the discussion.

Frogville Construction Workers Union #6000

The construction workers support the project because it will employ them locally and close to their families. The construction would also take several years and provide a long-term source of income.

Frogville Conservation Coalition

The coalition members are concerned citizens and local town residents who believe the new housing will destroy the habitat of the state's endangered northern cricket frog. They have evidence that the frog travels more than 1,000 feet from the lake to feed and would have to cross roads created for the senior housing.

Froggy Environmental Consultants

This company has been hired by the Senior Housing Corporation to study the site of the proposed senior housing and create a habitat investigation report. This report would make recommendations that help get the housing built but also protect the frogs and other plants and animals living in the wetland.

Frogville Elementary School Environmental Club

This group of students has a love of the environment and does not want to see their school mascot, the northern cricket frog, become extinct in the state. If there is a way to build the housing and protect the frogs, they would be very happy.

3. Students should be assigned to the following eight stakeholder groups that will be represented at a mock town meeting: Senior Housing Corporation, Senior Citizens Group, Frogville Planning Board, Frogville Chamber of Commerce, Frogville Construction Workers Union #6000, Frogville Conservation Coalition, Froggy Environmental Consultants, and Frogville Elementary School Environmental Club. Each group should

work together to read and discuss its description card.

4. Each group will receive or collect resources and props to prepare and make a presentation to the town council based on the point of view of its stakeholder description card.

5. Groups are provided time to discuss and research the situation statement among themselves and decide how the group would like to respond at the town meeting. Groups may create skits, prepare visuals such as graphs and sketches, and elect a spokesperson to represent their interests.

6. Each group will present its position to the town council. The council may ask questions of the group if clarification of the perspective is needed.

7. If time allows, groups may present questions to one another about their presentations.

8. The facilitator or invited representatives that make up a mock town council may then make a decision or recommendation of a solution that they believe is best for everyone.

Reflect and Explain

- What are the perspectives of each group represented at the town meeting?
- Where do the perspectives of each group cause conflict?
- What are the advantages and disadvantages of building the senior housing?
- Was there cooperation within your group? How did this help or hinder an outcome?
- What solution would you propose now that you've listened to the presentations?

Extensions

- Find an environmental issue to study and determine the different perspectives.
- Invite a local land-use planner or town official to discuss with the class how he or she works to prevent or solve conflict.

Resource Information

Flemming, D. 2000. *Where once there was a wood*. New York: Henry Holt.

Lesson 20:
Amphibian Art

Objectives
Students will create art that depicts amphibians. Students will research and write or present about the amphibian they chose to depict.

Method
Students create art based on amphibian species and analyze how art can impact culture and how artists use issues as subjects for their work.

Materials
Amphibian images, acrylic paints or art supplies to be determined by the instructor, field guides, reference books and websites
Safety note: Review Material Sheets Data Safety for acrylic paints.

Background Information
The use of art can help bring attention to causes or concerns when sensitive issues are displayed in a creative form. Creativity can help catch the eye and make people aware of issues they would not otherwise consider. The use of art can cause a variety of emotions in people, and creating art that represents local amphibians or **threatened** and

Grade Level: 2–4

Subject Area: science, art, language arts, history

Skills: analysis, application

Setting: inside

Lesson Duration: 90 minutes

Group Size: 15+

National Science Education Standards, Grades K–4:
- **Life Science:** The characteristics of organisms
- **Life Science:** Organisms and environments
- **Science in Personal and Social Perspectives:** Characteristics and changes in populations
- **Science in Personal and Social Perspectives:** Changes in environments
- **History of Science:** Science as a human endeavor

endangered amphibians can help bring attention to their existence and lead to discussion about their future. The decline in amphibians is caused by changes in the environment from habitat destruction, climate change, and invasive species to pesticide use, fire, disease, road mortality, and even harvest. These factors can lead to amphibians being listed as threatened species and eventually endangered species if populations are not protected or the causes of decline are not reversed. Students can use art to demonstrate the importance of amphibians in our ecosystems and culture and let the viewing community recognize that there is still time to protect these sensitive species.

Procedure

1. Ask students to pick an amphibian that they will promote. You may choose to limit choices to those amphibians that are threatened and endangered according to your state department of environmental protection or to those amphibians that are native to your local habitats.
2. Students identify things that belong in the habitat of the amphibian they choose and paint them into the background of their work. Include such things as lily pads or cattails for frogs, rotting logs for salamanders, or tunnels underground for caecilians.
3. When the background habitat is complete, students are instructed to paint their amphibian within the habitat created.
4. Student research should accompany the artwork with information that includes answers to questions such as "How does it protect itself?" "What is its habitat?" and "What does it eat?"

Reflect and Explain

- Is the subject represented in an accurate habitat?
- Why should we paint threatened and endangered animals? Will your artwork affect people? How so?
- Why did you choose to paint the amphibian they way you did? What were you trying to convey?
- Does your background match the habitat of the amphibian you painted? Explain.

Extensions

- Use a variety of art forms to depict and bring attention to wildlife.
- Depict a variety of species found at a frog pond and display your work as a large-scale field guide to the wildlife found in your local pond.
- Photograph student artwork and create a field guide or calendar that includes research or interesting facts students found about their chosen species.

- Display work in a prominent location within a school, nature center, museum, zoo, or town hall. Include an opening reception to introduce friends, families, and visitors to the art and communicate the message about current threats amphibians face.

Resource Information

Dobson, D., and J. Needham. 1997. *Can we save them? Endangered species of North America*. Watertown, MA: Charlesbridge.

Mackay, R. 2008. *The atlas of endangered species*. Berkeley: University of California Press.

National Wildlife Federation. 1997. *Endangered species: Wild and rare*. New York: McGraw-Hill.

Resource List

Children's Books

Bakken, A. 2006. *Uncover a frog*. Berkeley, CA: Silver Dolphin Books.

Barrett George, L. 1996. *Around the pond: Who's been here?* New York: Green Willow Books.

Baylor, B. 1995. *I'm in charge of celebrations*. New York: Aladdin Paperbacks.

Baylor, B. 1995. *The other way to listen*. New York: Aladdin Paperbacks.

Clarke, B., and L. Buller. 2005. *Amphibian*. New York: DK Children.

Ellis, B. 2006. *The web at dragonfly pond*. Nevada City, CA: Dawn Publications.

Faulkner, K. 1996. *Wide mouthed frog*. New York: Dial.

Flemming, D. 2000. *Where once there was a wood*. New York: Henry Holt.

Flemming, D. 2007. *In the small, small pond*. New York: Henry Holt.

Florian, D. 2001. *Lizards, frogs, and polliwogs*. New York: Harcourt Brace.

Fredericks, A., and J. Dirubbio. 2005. *Near one cattail: Turtles, logs and leaping frogs* Nevada City, CA: Dawn Publications.

French, V. 2008. *Growing frogs*. London: Walker Books.

Godwin, S. 1999. *The trouble with tadpoles: A first look at the life cycle of a frog*. London: Hodder Wayland.

Hawes, J. 2000. *Why frogs are wet*. New York: Harper Collins.

Heinz, B. 2000. *Butternut hollow pond*. New York: First Avenue Editions.

Heller, R. 1995. *How to hide a meadow frog and other amphibians*. New York: Penguin Group.

Hibbert, A. 1999. *A freshwater pond*. New York: Crabtree.

Himmelman, J. 1998. *A salamander's life*. Danbury, CT: Children's Press.

Himmelman, J. 1999. *A woodfrog's life*. Danbury, CT: Children's Press.

Kalman, B. 2000. *What is an amphibian?* New York: Crabtree.

Kaufman, B. 2006. *The life cycle of a frog*. New York: Crabtree.

Kent, J. 1982. *The caterpillar and the pollywog*. New York: Aladdin Paperbacks.

Lamstein, S. 2010. *Big night for salamanders*. Honesdale, PA: Boyd Mills Press.

Lescroart, J. 2008. *Icky sticky frog*. Atlanta, GA: Piggy Toes Press.

Lionni, L. 1996. *It's mine*. New York: Dragonfly Books.

Mazer, A. 1994. *The salamander room*. New York: Dragonfly Books.

McKinney, B. 2000. *Pass the energy, please*. Nevada City, CA: Dawn Publications.

Moignot, D. 1997. *Frogs (first discovery book)*. New York: Scholastic.

Morrison, G. 2002. *Pond*. New York: Houghton Mifflin.

Pallotta, J., and R. Masiello. 1990. *The frog alphabet book*. Watertown, MA: Charlesbridge.

Pfeffer, W. 1994. *From tadpole to frog*. New York: Harper Collins.

Pratt-Serafini, K. J. 2000. *Salamander rain:, A lake and pond journal*. Nevada City, CA: Dawn Publications.

Sidman, J. 2005. *Song of the water boatman*. Boston: Houghton Mifflin.

Silver, D., and P. Wynne.1997. *One small square: Swamp*. New York: McGraw-Hill.

Stewart, M. 2009. *Under the snow*. Atlanta, GA: Peachtree Publishers.

Tagholm, S., and B. Kitchen. 2000. *Animal lives: Frog*. Boston: Kingfisher.

Winer, Y. 2003. *Frogs sing songs*. Watertown, MA: Charlesbridge.

Field Guides

Conant, R., and J. T. Collins. 1998. *Peterson field guide to reptiles and amphibians, Eastern and Central North America*. 3rd ed. Boston: Houghton Mifflin.

Conant, R., R. Stebbins, and J. Collins. 1991. *Peterson first guide to reptiles and amphibians*. Boston: Houghton Mifflin.

Kenney, L., and M. Burne. 2001. *A field guide to the animals of vernal pools*. Boston: Massachusetts Division of Fisheries and Wildlife.

Mattison, C. 2007. *300 frogs: A visual reference to frogs and toads from around the world*. Ontario, Canada: Firefly Books.

Reid, G., S. Kaicher, and T. Dolan. 2001. *Pond life*. New York: Golden Guides, St. Martin's Press.

Zim, H., H. Smith, and J. Gordon. 2001. *Reptiles and amphibians*. New York: Golden Guides, St. Martin's Press.

Audio Aids

Davidson, C. 1999. *Frog and Toad Calls of the Rocky Mountains* CD. New York: Cornell Lab of Ornithology.

Eliot, L. 2004. *The Calls of Frogs and Toads* CD. Mechanicsburg, PA: Stackpole Books.

United States Geological Survey (Frog Quizzes). *Rwww.pwrc.usgs.gov/frogquiz/index.cfm*.

Video Aids

Deban, S. 2009. *Amphibian feeding movies*. University of South Florida. *http://autodax.net/feedingmovieindex.html*

Ford, S. (Producer). 2008. *Life in Cold Blood* DVD. United States: Warner Home Video.

Franke, V. (Producer). 2008. *Rattlers, Peepers, and Snappers*. United States. Peregrine Productions.

Ravenswood Media, Inc. 2009. *Why Frogs Call and Why We Should Listen*. DVD United States: Ravenswood Media.

Creating Amphibian Habitats

Biebighauser, T. R. 2000. *A guide to creating vernal ponds.* Washington DC: USDA Forest Service.

Ripple, K. L., and E. W. Garbisch. 2000. *POW! The planning of wetlands: An educator's guide.* St. Michaels, MD: Environmental Concern.

Wyzga, M. 1998. *Homes for wildlife: A planning guide for habitat enhancement on school grounds.* Concord, NH: New Hampshire Fish and Game Department.

Additional Educator Resources

Carroll, D. 1999. *Swamp walker's journal: A wetlands year.* Boston: Houghton Mifflin.

Dobson, D., and J. Needham. 1997. *Can we save them? Endangered species of North America.* Watertown, MA: Charlesbridge.

Lannoo, M. 2008. *Malformed frogs: The collapse of aquatic ecosystems.* Berkeley: University of California Press.

Mackay, R. 2008. *The atlas of endangered species.* Berkeley: University of California Press.

Mendelson, J. 2009. Considerations and recommendations for raising live amphibians in classrooms. Society for the Study of Amphibians and Reptiles. *www.ssarherps.org/documents/amphibians_in_classroom.pdf.*

National Science Teachers Association. Responsible use of live animals and dissection in the science classroom. *www.nsta.org/about/positions/animals.aspx.*

National Wildlife Federation. 1997. *Endangered species: Wild and rare.* New York: McGraw-Hill.

Phillips, K. 1994. *Tracking the vanishing frogs: An ecological mystery.* New York: St. Martin's Press.

Ryan, M. 2001. *Anuran communication.* Washington, DC: Smithsonian Press.

Schneider, R. L., M. E. Krasny, and S. J. Morreale. 2001. *Hands-on herpetology: Exploring ecology and conservation.* Arlington, VA: NSTA Press.

Stebbins, R. C., and N. W. Cohen. 1995. *A natural history of amphibians.* Princeton, NJ: Princeton University Press.

Temko, F. 1986. *Paper pandas and jumping frogs.* San Francisco: China Books.

West, L., and W. P. Leonard. 1997. *How to photograph reptiles and amphibians.* Mechanicsburg, PA: Stackpole Books.

Scientific Equipment Suppliers

Acorn Naturalists: *www.acornnaturalists.com*
Carolina Biological Supply: *www.carolina.com*
Home Science Tools: *www.homesciencetools.com*

Related Web Resources

A Thousand Friends of Frogs: *www.hamline.edu/cgee/frogs*
This site connects children, parents, educators, and scientists to study and celebrate frogs and their habitats. It offers good classroom resources and lessons focusing on frogs.

Resource List

Amphibian Specialist Group: *www.amphibians.org*

The Amphibian Specialist Group strives to conserve biological diversity by stimulating, developing, and executing practical programs to conserve amphibians and their habitats around the world. This will be achieved by supporting a global web of partners to develop funding, capacity, and technology transfer to achieve shared strategic amphibian conservation goals.

National Wildlife Federation, Frog Watch USA: *www.aza.org/frogwatch*

This volunteer monitoring program collects data on both frogs and toads to create awareness of amphibian decline.

North American Reporting Center for Amphibian Malformations: *www.nbii.gov*

This organization accepts reports on amphibian malformations. The reports provide an important baseline of data on the health and fitness of existing amphibian populations; patterns of reported malformation occurrences can help direct further research and study of these phenomena ito determine causes.

Partners in Amphibian and Reptile Conservation: *www.parcplace.org*

This organization is a wide-ranging partnership dedicated to the conservation of the herpetofauna— reptiles and amphibians—and their habitats. The primary concern is with habitat protection and endangered and threatened species, as well as keeping common species common.

Save The Frogs: *www.savethefrogs.com*

Save The Frogs is America's first and only public charity dedicated exclusively to amphibian conservation. Their mission is to protect amphibian populations and to promote a society that respects and appreciates nature and wildlife.

Society for the Study of Amphibians and Reptiles: *www.ssarherps.org*

This international nonprofit—established to advance research, conservation, and education concerning amphibians and reptiles—was founded in 1958. They also support numerous grant programs, international exchanges, and international cooperative efforts for study and publications.

Tree Walkers International: *www.treewalkers.org*

This organization supports the protection, conservation, and restoration of wild amphibian populations through hands-on action both locally and internationally. You can also learn about creating your own frog pond through their operation frog pond program.

USA National Phenology Network: *www.usanpn.org*

The USA National Phenology Network brings together citizen-scientists, government agencies, nonprofit groups, educators, and students of all ages to monitor the effects of climate change on plants and animals in the United States.

U.S. Deptartment of Transportation: Federal Highway Administration Critter Crossing: *www.fhwa.dot. gov/environment/wildlifecrossings/main.htm*

This governmental organization includes information about protecting wildlife with critter crossings for salamanders and other animals in an effort to link habitats and reduce road kill.

Vernal Pool Association: *www.vernalpool.org*

The Vernal Pool Association began in 1990 as an environmental outreach project at Reading Memorial High School in Reading, Massachussetts. It is now an independent group of individuals attempting to educate others about vernal pool ecology, the local environment, biodiversity, and the protection of our resources.

Glossary

Abiotic: Never-living thing such as a rock, sunlight, water, or air.

Adult: A biological grown or mature animal.

Amphibian: Class of cold-blooded animals that metamorphose from a juvenile water-breathing form to an adult air-breathing form. Frogs, toads, salamanders, newts, and Caecilians are all included.

Amplexus: When amphibians grasp as part of the mating process.

Anura: Order of tail-less amphibians that includes frogs and toads.

Aquatic: Animals that live in water.

Arboreal: Animals that live in trees.

Army of frogs: A group of frogs.

Bask: To be exposed to warmth, especially from the Sun.

Behavior: The actions of an organism in relation to the environment.

Biodiversity: All the kinds of plants and animals within a habitat or the world.

Bio-indicator: An organism whose health helps us assess the health of an ecosystem.

Biotic: Living thing such as plants or animals.

Caecilian: A limbless amphibian that resembles an earthworm or snake. They are only found in tropical areas of the world.

Camouflage: Body coloration pattern that helps hide an animal within a habitat.

Captivity: Being confined to a space and unable to escape.

Carnivore: An organism that eats only animals.

Chytrid fungi: An infectious disease of amphibians that has led to population declines.

Citizen-scientist: Individual volunteers with no specific scientific training who perform research-related tasks.

Classification: To arrange things according to their similarities.

Cold-Blooded: See **Ectothermic**.

Conservation: The protection and management of biodiversity.

Consumer: An animal that eats other animals or plants.

Decomposer: An organism that eats dead or decaying organisms and, in doing so, carries out the natural process of decomposition.

Deformity: A mutation such as a frog with three legs.

Democracy: A system in which citizens elect people to represent them.

Disturbance: A change in environmental conditions that alters an ecosystem.

Diurnal: Active during the daylight.

Diversity: The variation of life forms within a given ecosystem.

Ecology: Interactions between organisms and their environment.

Ecosystem: An area where living and nonliving things interact.

Ectothermic: Animals that rely primarily on external sources of heat to regulate their body temperature. For example, amphibians may bask in the Sun.

Egg: The first stage of metamorphosis in amphibians.

Endangered species: Population of animals that are at risk of becoming extinct.

Endoskeleton: A skeleton on the inside of the body.

Extinction: The end of an organism or the death of the last individual of a species.

Food chain: The flow of energy from organism to organism.

Food web: The predator-prey relationships within an ecosystem.

Frog: A tail-less amphibian, often with long hind legs, a short body, webbed fingers and toes, and protruding eyes.

Gills: A respiratory organ in amphibians that allows them to gather oxygen from water. The gills are usually replaced by lungs as the amphibian metamorphoses into an adult.

Habitat: A place where a plant or animal lives with food, water, shelter, and space.

Habitat destruction: When the health of a habitat is lowered.

Habitat fragmentation: When a habitat is broken into small, nonconnected sections.

Herbivore: An organism that eats only plants.

Herpetologist: A person who studies reptiles and amphibians.

Herpetology: The study of reptiles and amphibians.

Hibernation: The time when an animal's body functions slow down.

Indicator: An organism whose presence may inform us about the health of a habitat.

Interconnected: Living and nonliving things connected with one another in an ecosystem.

Invertebrate: An animal that has no backbone.

Lake: A body of water larger than a pond.

Larva: A juvenile form amphibians and some animals undergo before metamorphosis into adults. Often called **tadpole** when referring to a frog.

Life cycle: The stages in an animal's life from egg to adult.

Lung: Organ that allow adult amphibians to gather oxygen from air.

Macroinvertebrates: An invertebrate large enough to be seen without using a microscope.

Metamorphosis: The process of an amphibian or animal changing after birth into an adult form.

Migration: The movement of an animal from one habitat to another for breeding, hibernation, or feeding.

Nocturnal: Active during the night.

Omnivore: An organism that eats both plants and animals.

Organism: Any living system such as an animal, plant, or fungus.

Parotid gland: An external skin gland on the back, neck, and shoulder of some amphibians that secretes a milky alkaloid substance to deter predators.

Permeable: Having pores that allow water and other substances to pass through. Amphibians have permeable skin.

Phenology: The study of recurring plant and animal life cycle stages.

Poison Dart Frog: A group of frogs from Central and South America with bright colors and a somewhat toxic level of secretions from their skin.

Polliwog: See **tadpole**.

Pollutant: A foreign substance that makes something impure or unhealthy.

Pond: A body of water smaller than a lake.

Posture: The way an animal holds and positions its body to expose certain colors or parts.

Predator: An organism that is hunting prey.

Prey: An organism that is attacked by a predator.

Producer: Organisms such as plants on land or algae in water that make food by photosynthesis.

Reptile: Class of cold-blooded vertebrate that has scaly skin and leathery eggs. Examples include turtles, snakes, lizards, and alligators.

Salamander: A slender-bodied amphibian with both front and rear legs, a short nose, and a long tail.

Secretion: A fluid that flows out of skin pores onto the skin.

Spawn: Eggs of amphibians or fish.

Species: A group of organisms that have the same traits and can reproduce.

Stakeholders: A person or group affected by the actions of others.

Swamp: A wetland with temporary or permanent flooding of water.

Tadpole: The aquatic larval stage of an amphibian, particularly of a frog or toad.

Terrestrial: To live on land.

Tetrapod: A four-limbed animal.

Threatened: Any species vulnerable to extinction.

Toad: An amphibian similar to a frog but with leathery skin and warts.

Torpor: Temporary hibernation or a short-term state of decreased activity.

Toxic: Relating to damaging or harmful chemicals that can cause illness or death.

Tympanic membrane: Also known as the eardrum.

Vernal pool: A temporary pool of water.

Vertebrae: A long column of bones.

Vertebrate: An animal that has a backbone or a spine.

Vocalization: Animal communication involving the use of vocal cords.

North American Association for Environmental Education Guidelines Alignment Chart

Amphibian Curriculum Guide

Lesson Correlations to North American Association for Environmental Education Guidelines, Grades K–4

Lesson

Strand 1: Questioning, Analysis, and Interpretation Skills	1	2	3	4	5	6	7	8	9	10	11	12	13	14	15	16	17	18	19	20
A. Generate and develop questions that are appropriate for initiating inquiry.								•	•	•		•	•				•	•	•	
B. Design simple investigations.										•							•			
C. Locate and collect information about the environment and environmental topics from a variety of resources.		•		•				•	•								•	•	•	
D. Understand the need to use reliable information; explain some of the factors to consider in judging the merits of the information they are using.				•						•	•				•				•	
E. Describe data and organize information to show relationships and patterns.	•	•	•	•				•	•	•	•	•		•	•	•				
F. Work with models and simulations, using them to describe relationships, patterns, and processes.	•	•	•		•				•											
G. Describe their observations and develop simple explanations.	•	•		•	•	•			•	•	•	•		•	•	•				

Lesson

	1	2	3	4	5	6	7	8	9	10	11	12	13	14	15	16	17	18	19	20
Strand 2: Knowledge of Environmental Processes and Systems																				
2.1—The Earth as a Physical System																				
A. Identify and explain changes and differences in the physical environment.										●						●				
B. Identify and describe basic characteristics of and changes in matter.																				
C. Describe the basic sources and uses of some different forms of energy (light, heat, etc.).									●											
Strand 2: Knowledge of Environmental Processes and Systems																				
2.2—The Living Environment																				
A. Identify similarities and differences among a wide variety of living organisms; describe organisms' basic needs, habitats, and ways organisms meet their needs in different habitats.								●	●											
B. Explain that both plants and animals have different characteristics and that many of the characteristics are inherited from their parents.	●																			
C. Explain basic ways in which organisms are related to their environments and to other organisms.		●				●	●		●	●		●				●		●	●	
D. Explain that living things need some source of "energy" to live and grow and that matter is recycled—e.g., through life, growth, death, and decay.		●							●											

Lesson

Strand 2: Knowledge of Environmental Processes and Systems	1	2	3	4	5	6	7	8	9	10	11	12	13	14	15	16	17	18	19	20
2.3—Humans and Their Societies																				
A. Identify ways that people act as individuals and as group members, and give examples of ways groups influence individual actions.															•		•	•	•	
B. Give examples of how experiences and places may be interpreted differently by people with different cultural backgrounds, at different times, or with other frames of reference.															•	•	•	•	•	•
C. Describe government and economic systems that exist because people living together in groups need ways to do things (such as provide for needs and wants, maintain order, and manage conflict).																	•		•	
D. Understand how people are connected at many levels—including the global level—by actions and common responsibilities that concern the environment.																	•		•	
E. Recognize that change is a normal part of individual and societal life and that conflict is rooted in different points of view.																			•	•

Strand 2: Knowledge of Environmental Processes and Systems

2.4—Environment and Society

	1	2	3	4	5	6	7	8	9	10	11	12	13	14	15	16	17	18	19	20
A. Identify ways people depend on, change, and are affected by the environment.							•			•						•			•	•
B. Describe ways places differ in their physical and human characteristics.																	•		•	
C. Demonstrate an understanding of "resources" and describe various sources and origins of resources they use in their lives.									•								•			
D. Understand that technology is an integral part of human existence and culture.							•										•		•	
E. Identify and describe a range of local environmental issues and understand that people in other places also experience environmental issues.															•			•	•	

Strand 3: Skills for Understanding and Addressing Environmental Issues

3.1—Skills for Analyzing and Investigating Environmental Issues

	1	2	3	4	5	6	7	8	9	10	11	12	13	14	15	16	17	18	19	20
A. Identify and investigate local environmental issues.										•					•		•	•	•	•
B. Speculate about and explore the social, economic, and environmental consequences of issues and proposed solutions to them.										•								•	•	
C. Identify and evaluate alternative approaches to resolving issues.																		•	•	
D. Discuss and critique ideas representing different perspectives; hear and respect viewpoints that differ from their own.																		•	•	•

Lesson

Strand 3: Skills for Understanding and Addressing Environmental Issues

3.2—Decision-Making and Citizenship Skills

	1	2	3	4	5	6	7	8	9	10	11	12	13	14	15	16	17	18	19	20
A. Examine and express their own views on environmental issues.																		●	●	●
B. Consider whether they believe action is needed in particular situations and whether they think they should be involved.							●			●							●	●	●	
C. Learn the basics of individual and collective action, by participating in close-to-home issues of their choosing.																	●	●	●	
D. Evaluate the results of actions, understanding that civic actions have consequences.																		●	●	

Strand 4: Personal and Civic Responsibility

	1	2	3	4	5	6	7	8	9	10	11	12	13	14	15	16	17	18	19	20
A. Identify the fundamental principles of U.S. society and explain their importance in the context of environmental issues.																			●	
B. Understand the basic rights and responsibilities of citizenship.																			●	
C. Possess a realistic self-confidence in their effectiveness as citizens.																	●	●	●	
D. Understand that they have responsibility for the effects of their actions.																		●	●	

Index